Advances in Intelligent and Soft Computing

157

Editor-in-Chief

Prof. Janusz Kacprzyk
Systems Research Institute
Polish Academy of Sciences
ul. Newelska 6
01-447 Warsaw
Poland
E-mail: kacprzyk@ibspan.waw.pl

For further volumes:
http://www.springer.com/series/4240

Juan M. Corchado Rodríguez, Javier Bajo Pérez,
Paulina Golinska, Sylvain Giroux,
and Rafael Corchuelo (Eds.)

Trends in Practical Applications of Agents and Multiagent Systems

10th International Conference on Practical
Applications of Agents and Multi-Agent
Systems

 Springer

Editors
Juan M. Corchado Rodríguez
Departamento de Informática y Automática
Facultad de Ciencias
Universidad de Salamanca
Salamanca
Spain

Javier Bajo Pérez
Escuela Universitaria de Informática
Universidad Pontificia de Salamanca
Salamanca
Spain

Paulina Golinska
Poznan University of Technology
Insitute of Manangement Engineering
Poznan
Poland

Sylvain Giroux
Département de mathématiques
et d'informatique
Faculté des Sciences
Université de Sherbrooke
Canada

Rafael Corchuelo
ETSI Informática
Sevilla
Spain

ISSN 1867-5662   e-ISSN 1867-5670
ISBN 978-3-642-28794-7   e-ISBN 978-3-642-28795-4
DOI 10.1007/978-3-642-28795-4
Springer Heidelberg New York Dordrecht London

Library of Congress Control Number: 2012933090

Printed on acid-free paper

Springer is part of Springer Science+Business Media (www.springer.com)

Preface

PAAMS'12 Workshops complement the regular program and the special sessions with new or emerging trends of particular interest connected to multi-agent systems.

PAAMS, the International Conference on Practical Applications of Agents and Multi-Agent Systems is an evolution of the International Workshop on Practical Applications of Agents and Multi-Agent Systems. PAAMS is an international yearly tribune to present, to discuss, and to disseminate the latest developments and the most important outcomes related to real-world applications. It provides a unique opportunity to bring multi-disciplinary experts, academics and practitioners together to exchange their experience in the development of Agents and Multi-Agent Systems.

This volume presents the papers that have been accepted for the 2012 in the workshops: Workshop on Agents for Ambient Assisted Living, Workshop on Agent-Based Solutions for Manufacturing and Supply Chain, Workshop on Agents and Multi-agent systems for Enterprise Integration.

We would like to thank all the contributing authors, the sponsors (IEEE Systems Man and Cybernetics Society Spain, AEPIA Asociación Española para la Inteligencia Artificial, APPIA Associação Portuguesa Para a Inteligência Artificial, CNRS Centre national de la recherche scientifique), as well as the members of the Program Committees of the workshops and the Organizing Committee for their hard and highly valuable work. Their work has helped to contribute to the success of the PAAMS'12 event. Thanks for your help, PAAMS'12 would not exist without your contribution.

Juan M. Corchado Rodríguez
Javier Bajo
PAAMS'12 Organizing Co-chairs

Organization

Workshops

W1 – Workshop on Agents for Ambient Assisted Living
W2 – Workshop on Agent-Based Solutions for Manufacturing and Supply Chain
W3 – Workshop on Agents and Multi-agent systems for Enterprise Integration

Workshop on Agents for Ambient Assisted Living Committee

Kasper Hallenborg (Co-chairman)	University of Southern Denmark, Denmark
Sylvain Giroux (Co-chairman)	Université de Sherbrooke, Canada
Bruno Bouchard	Université Paris-Dauphine, France
Fabrice Peyrard	Toulouse II University, France
Nikolaos Spanoudakis	Technical University of Crete, Greece
Olivier Boissier	ENS Mines de Saint-Etienne, France
Patrice Roy	Unversity of Sherbrooke, Canada
Pierre Busnel	Unversity of Sherbrooke, Canada
Valerie Camps	University of Toulouse, France
Abdenour Bouzouane	University of Québec, Canada
Charles Gouin-Vallerand	Universirty of Sherbrooke, Canada
Nadjia Kara	University of Québec, Canada
Pierre Rumeau	University of Toulouse, France
Dante Tapia	University of Salamanca, Spain
Carolina Zato	University of Salamanca, Spain

Workshop on Agent-Based Solutions for Manufacturing and Supply Chain Committee

Pawel Pawlewski (Co-chairman)	Poznan University of Technology, Poland
Paulina Golinska (Co-chairman)	Poznan University of Technology, Poland
Paul-Eric Dossou	ICAM Vendee, France
Zbigniew J. Pasek	IMSE/University of Windsor, Canada

Arkadiusz Kawa   Poznan University of Economy, Poland
Grzegorz Bocewicz   Koszalin University of Technology, Poland
F. Javier Otamendi   Universidad Rey Juan Carlos, Spain

Workshop on Agents and Multi-agent systems for Enterprise Integration Committee

Organizing Committee

José L. Álvarez   Universidad de Huelva, Spain
José L. Arjona   Universidad de Huelva, Spain
Fernando Bellas   Universidad de la Coruña, Spain
Rafael Corchuelo   Universidad de Sevilla, Spain
Javier López   Universidad de la Coruña, Spain
Paula Montoto   Universidad de la Coruña, Spain
Alberto Pan   Universidad de la Coruña, Spain
Hassan A. Sleiman   Universidad de Sevilla, Spain
Inma Hernández   Universidad de Sevilla, Spain

Scientific Committee

Albert Cheng   University of Houston, USA
Antonio Dourado   Universidade de Coimbra, Portugal
Carlos R. Rivero   Universidad de Sevilla, Spain
Christoph Bussler   Saba Software, Inc., USA
Claus Pahl   Dublin City University, Ireland
Daniel Lemire   UQAM - University of Quebec at Montreal,
                                  Canada
David Ruiz   Universidad de Sevilla, Spain
Dimitris Karagiannis   University of Vienna, Austria
Enrique Menor   Intelligent Integration Factory, S.L., Spain
François Vernadat   Cour des Comptes Européene, Luxembourg
Gracja Wydmuch   Wroclaw University of Economics, Poland
Gustavo Rossi   Universidad Nacional de La Plata,
                                  Argentina
Hatem Ben Sta   Tunisia University, Tunisia
Hichem Omrani   CEPS/INSTEAD, Luxembourg
Janis Grundspenkis   Riga Technical University, Latvia
Joseph Giampapa   Carnegie Mellon University, USA

Juan Antonio Garrido	Intelligent Integration Factory, S.L., Spain
Jun Hong	Queen's University Belfast, UK
Kauko Leiviskä	University of Oulu, Finland
Luis M. Camarinha-Matos	New University of Lisbon, Portugal
Maiga Chang	Athabasca University, Canada
Maki Habib	The American University in Cairo, Egypt
Marcin Paprzycki	Instytut Slawistyki, Poland
Matthias Nickles	Technical University of Munich, Germany
Michael Vassilakopoulos	University of Central Greece, Greece
Ozgur K. Sahingoz	Turkish Air Force Academy, Turkey
Sadok Ben Yahia	Faculté des Sciences de Tunis, Tunisia
Samia Oussena	Thames Valley University, UK
Schahram Dustdar	Vienna University of Technology, Austria
Sebastián Ventura Soto	University of Cordoba, Spain
Viacheslav Wolfengagen	Institute JurInfoR, Russian Federation
Vijay Sugumaran	Oakland University, USA
Xiangfeng Luo	Shanghai University, China

Contents

Workshop on Agents for Ambient Assisted Living (AAAL'12)

Applying Model-Based Techniques to the Development of UIs for Agent Systems 1
Sebastian Ahrndt, Dirk Roscher, Marco Lützenberger, Andreas Rieger, Sahin Albayrak

Virtual Agents in Next Generation Interactive Homes 9
Rafael Del-Hoyo, Luis Sanagustín, Carolina Benito, Isabelle Hupont, David Abadía

Accessing Cloud Services through BDI Agents Case Study: An Agent-Based Personal Trainer to COPD Patients 19
Kasper Hallenborg, Pedro Valente, Yves Demazeau

Evaluating the n-Core Polaris Real-Time Locating System in an Indoor Environment 29
Dante I. Tapia, Óscar García, Ricardo S. Alonso, Fabio Guevara, Jorge Catalina, Raúl A. Bravo, Juan Manuel Corchado

Workshop on Agent-Based Solutions for Manufacturing and Supply Chain (AMSC'12)

Cyclic Scheduling for Supply Chain Network 39
Grzegorz Bocewicz, Robert Wójcik, Zbigniew Banaszak

Improving Production in Small and Medium Enterprises 49
María L. Borrajo, Javier Bajo, Juan F. De Paz

Multiagent System for Detecting and Solving Design-Time Conflicts in Civil Infrastructure 57
Jaume Domínguez Faus, Francisco Grimaldo, Fernando Barber

Simulation and Analysis of Virtual Organizations of Agents 65
Elena García, Virginia Gallego, Sara Rodríguez, Carolina Zato,
Juan F. de Paz, Juan Manuel Corchado

**Using Simulation Based on Agents (ABS) and DES in Enterprise
Integration Modelling Concepts** 75
Pawel Pawlewski, Paul-Eric Dossou, Paulina Golinska

**A Genetic Algorithm-Based Heuristic for Part-Feeding Mobile Robot
Scheduling Problem** 85
Quang-Vinh Dang, Izabela Ewa Nielsen, Grzegorz Bocewicz

Workshop on Agents and Multi-Agent Systems for Enterprise Integration (ZOCO'12)

ACA Multiagent System for Satellite Image Classification 93
Moisés Espínola, José A. Piedra, Rosa Ayala, Luís Iribarne,
Saturnino Leguizamón, Massimo Menenti

**Automatic Extraction of Geographic Locations on Articles of Digital
Newspapers** 101
Cesar García Gómez, Ana Flores Cuadrado, Jorge Díez Mínguez,
Eduardo Villoslada de la Torre

An Experiment to Test URL Features for Web Page Classification 109
Inma Hernández, Carlos R. Rivero, David Ruiz, José Luis Arjona

On Relational Learning for Information Extraction 117
Patricia Jiménez, José Luis Arjona, Jose Luis Álvarez

Automatic Optimization of Web Navigation Sequences 125
José Losada, Juan Raposo, Alberto Pan, Javier López

Metabolic Pathway Data and Application Integration 133
Ismael Navas-Delgado, Maria Jesús García-Godoy,
José F. Aldana-Montes

**Analysing the Effectiveness of Crawlers on the Client-Side Hidden
Web** 141
Víctor M. Prieto, Manuel Álvarez, Rafael López-García, Fidel Cacheda

Information Extraction Framework 149
Hassan A. Sleiman, Rafael Corchuelo

Behavior Pattern Simulation of Freelance Marketplace 157
Vadim Zuravlyov, Anton Matrosov, Dmitrijs Rutko

Author Index 165

Applying Model-Based Techniques to the Development of UIs for Agent Systems

Sebastian Ahrndt*, Dirk Roscher, Marco Lützenberger, Andreas Rieger, and Sahin Albayrak

Abstract. To counter difficulties of user interface (UI) development, model based techniques became firmly established over the last years. The basic idea of model based user interface development (MBUID) is to formally specify a UIs appearance and behaviour by means of several models. Especially for distributed multi-agent systems, the appliance of MBUID can be most promising. Agent applications involve many different execution platforms and heterogeneous devices and perfectly fit for Ambient Assisted Living landscapes due to their innate characteristics of distribution and autonomy. When it comes to agent systems, one always has to consider the fact that humans have to communicate with agents in the end. It is our opinion that most approaches neglect this fact and thus cut the dynamics and the capabilities of distributed multi-agent systems. Hence in this work, we present an approach for the development of UIs for software agents which applies model based techniques and also retains all degrees of freedom for the underlying multi-agent system.

1 Introduction

Ambient Assisted Living (AAL) is strongly facilitated by the vision of ubiquitous computing, where smart interacting devices are integrated into the everyday life. As a matter of fact, the importance of AAL services increases over time as a result of demographic changes. In order to maintain the quality of life – especially for the elderly – technologies are required which support a living at home in many aspects, such as autonomy, security and health.

Over the last years, *Agent Oriented Software Engineering* (AOSE) has evolved as suitable technique for the development of AAL systems [6]. The reason for this is that multi-agent systems (MAS) perfectly fit for AAL landscapes due to their

Sebastian Ahrndt · Dirk Roscher · Marco Lützenberger · Andreas Rieger · Sahin Albayrak
DAI-Labor, Technische Universität Berlin, Ernst-Reuter-Platz 7, 10587 Berlin, Germany
e-mail: sebastian.ahrndt@dai-labor.de
*Corresponding author.

J.M.C. Rodríguez et al. (Eds.): Trends in PAAMS, AISC 157, pp. 1–8.

innate characteristics of distribution and autonomy. In fact, agent-based systems are able to match many requirements of Ambient Assisted Living. However, it is generally agreed, that the success of software applications is not only founded by the capability of the application itself, but also by the quality of its handling and also by its usability. When it comes to AAL, one always has to consider, that the target audience is usually the elderly. This raises many challenges for software developers as elderly people are not as experienced in handling software as younger people are [7]. Further, the situation is aggravated, as different device types and many interaction modalities (such as voice-, mouse-, touch- and gesture-based interaction) have to be taken into account. To sum up, in order to support users in the spirit of AAL, developers have to provide users with intuitive and yet non-intrusive control mechanisms.

However, serving multi-modal interaction possibilities and also supporting different device types results in countless UI variations and even more configuration options. To counter this problem, user interfaces for similar application areas are frequently developed in compliance with the *Model-Based User Interface Development* (MBUID) paradigm. The basic idea of MBUID is to formally specify a user interface's appearance and behaviour by means of several models from which executable code can be derived. Further, interpreter-based *Model-based User Interfaces* (MBUI) have the ability to manipulate their models at runtime and to dynamically adjust to the current execution context.

In this paper, we argue that the combination of software agents and MBUIs is a sophisticated way to increase the comfortability when developing AAL services. We start with a survey on related approaches (see Section 2). Afterwards, we present an approach that enables the development of agent-applications with MBUIs that are interpreted at runtime in order to provide a holistic user experience for AAL environments (see Section 3). Subsequently, we will illustrate a proof-of-concept implementation of an AAL service which we currently present in the showroom of our research institute (see Section 4). We proceed by discussing practical experiences we have made thus far and finally wrap up with a conclusion (see Section 5).

2   Related Work

Prior to our development, we performed a survey on existing approaches. The HCI community provides an established body of works regarding MBUIs and MBUI development environments [11, 14]. However, these works do not contribute to the integration of MBUIs into the agent domain. As a matter of fact, this area of research is only sparely covered. The agent community for instance tries to counter the complexity of UI development by web-based solutions [1, 15]. As these approaches are not directly comparable with our architecture, we identified some others which are described next.

Braubach et al. [4], for instance, introduce *Vesuf*, a development environment for MBUIs. Vesuf was not streamlined for agent applications, however, the framework was tested in real life, in an urban hospital facility[1], where it demonstrated its capability to generate adaptable UIs for software agents. In their work, the authors emphasise the difficulties in developing UIs for agent systems and argue, that interpreter-based MBUIs are capable to overcome most of the mentioned problems.

Eisenstein and *Rich* [8] propose an architecture which is based on task-models and which facilitates the development of collaborative interface agents. The authors apply task-models to control the behaviour of agents and also as foundation for the UI. Development is done in compliance with the underlying task-model and supported by a set of editors, each one geared towards a specific part of the application.

Tran et al. [18] present an approach which applies MBUID for data systems. The authors present an agent-based framework, that allows for the automated generation of database UIs and application code, which is based on a combination of task-, context-, and domain model. As the different models have different roles, agents are used for the code generation as well.

Pruvost and *Bellik* [16] present a framework for multi-modal interaction in ubiquitous systems. The framework is a part of the *European ATRACO* project[2]. One interesting aspect of this work is that agents negotiate on how to render the MBUIs.

Braubach et al. [4] impressively demonstrate the capabilities of merging MBUIs with agent systems, although their approach was not intended for agent systems in the first place. As a result to this design decision, their architectural presentation model lacks depth. In our opinion, the used presentation model does not provide enough information. Hence, it has to be extended with modality-dependent informations, which leads to one UI descriptions for each supported modality. Further, although *Vesuf* is an interpreter-based MBUID environment, there is no context-model available. Hence, the UIs cannot adapt to the actual context-of-use at runtime. The other examinees focus on particular aspects and disregard the bigger picture of a holistic user experience. Nevertheless, *Pruvost* and *Bellik* present interesting ideas, which gives us visions for future extensions of our work, as for instance agents which negotiate about the most adequate way of interaction.

To sum up, our survey shows, that although there are many approaches to develop MBUIs, only a few of them have been applied and tested in conjunction with agent systems. Yet, it is our believe, that a multi-agent systems and MBUIs is a promising combination for AAL environments.

3 Approach

As mentioned above, the development of multi-modal UIs is a complex task. Interpreter-based MBUID can be used as it counters many difficulties and also suits well for the realm of AAL. Based on our survey we can state that current MBUI technology has, as yet, not found its way into the agent domain and vice versa. It is

---

[1] MedPAge (Medical Path Agents), see http://vsis-www.informatik.uni-hamburg.de/projects/medpage

[2] Adaptive and Trusted Ambient Ecologies, see www.uni-ulm.de/in/atraco

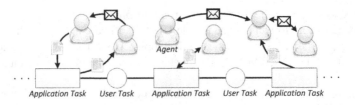

Application Task   User Task   Application Task   User Task   Application Task

Fig. 1 Abstract illustration of the approach, enabling agents to perform the application tasks of the task-model.

our objective to narrow the gap between both technologies and in the following we describe our way to achieve this goal. We start by outlining the target system and proceed by getting granular on our approach. Subsequently, we introduce applied technologies and finally, we argue on how the presented systems works together and fosters the interplay between MBUIs and multi-agent system technology.

3.1 AOSE meets MBUID

MBUI development applies several models in order to ensure device independence, multi-modal interaction and context-awareness. Each model encapsulates particular information on some part of the application as a whole. Runtime systems interpret these models and derive UIs which are optimised for a given execution context.

However, although there are several different models available, one is involved in the majority of MBUID environments – the task-model [5]. The task-model formalises the general workflow[3] of the application and distinct between tasks of the user and tasks that belong to the application's logic. Task-models can be described by using many languages, and reach from static to dynamic and executable ones.

Agents on the other hand are usually compelled to some application goal and manage the application's logic accordingly. In order to enable MBUIs for multi-agent systems we have to ensure that the application's tasks can be interpreted and performed by the agents. Figure 1 illustrates the principle.

In the example, the task-model is represented as a chain of application- and user tasks. Whenever the runtime system detects an application task, a referenced back-end service should be called – in our case an agent. Further, required data should be forwarded to the agent, yet, as MBUIs and multi-agent systems are usually based on different technologies and have different conflicting properties (straight definition vs. degrees of freedom), this task is not easily accomplished. In order to foster communication between MBUIs and multi-agent systems, we developed the *Human Agent Interface* [2] (HAI). HAI was designed to facilitate the integration of user interface technologies into agent applications. During runtime, HAI acts as a gateway between UIs and MAS, hiding particular UI details from the agent application and vice versa. Thus, to convert and deliver UI messages to the agent world and to forward responses from the agent application back to the user interfaces is HAI's

---

[3] A workflow is considered to be the tasks that can be reached.

main purpose. We designed HAI to be independent from any specific UI technology and also as extension to the Model-View-Controller (MVC) architecture. Due to its characteristics, HAI constitutes a suitable foundation for the problem we address in this work.

3.2   From Theory to Practice

Before presenting a system which takes AOSE and MBUID into account we want to provide a short outline of the applied technologies. Although HAI is not restricted to a particular agent framework, we frequently used HAI in combination with the *Java-based Intelligent Agents Componentware* [10] (JIAC V). JIAC V is a Java based agent framework which has been developed at the Technical University of Berlin since 1998. It combines service-oriented with agent-oriented concepts and offers conformity to FIPA standards[4].

Using model-based development to implement user interfaces provides many advantages. Nevertheless, as an objective of our work, we want to prevent UI- and agent developers from affecting each other. In order to do so, we applied the *Multi-Access Service Platform* [3] (MASP), as it allows model-based development and clearly distinct between UI and application. MASP task-models are based on the widely accepted *ConcurTaskTree* notation [13] (CTT). CTT separates task-models into four types of tasks: User tasks, application tasks, interaction tasks and abstract tasks. User and interaction tasks are performed by the user. Application tasks are executed by the system and abstract tasks are complex actions which can not be expressed by the other ones. In order to provide information on the execution sequence (e.g. parallel, step-by-step) and interdependencies between them, the tasks are ordered by means of LOTUS operators. Although CTT is a good foundation for user-centric design, it neglects some requirements for AAL sceneries and had to be extended in some aspects [9]. To start with, AAL environments – especially those with agents – continuously collect sensor data and may identify situations in which user interaction tasks have to be triggered or disabled. Classical CTT do not support this kind of behaviour and therefore prevents proactive agents to adjust the user interface to the latest set of environmental data.

Figure 2 illustrates the architecture of the implemented system including MASP, HAI and the MAS. Once an application task occurs, the additional backend service is executed. In order to assure that agents are able to manage the respective application tasks, we have implemented the *HAIService*, which manages the mapping process. After the HAIService was called a HAI service message is generated and send to the HAI (1). This message contains additional data (e.g. name, required capabilities or supported input/output parameters) and an identifier for the designated agent, which is used by HAI to establish a permanent connection to the responding agent for further UI requests. Subsequently, HAI converts the UI message into an agent message (in compliance with the FIPA ACL standard) and forwards the message to the agent system (2). The agent system now processes the incoming message

---

[4] FIPA – The Foundation for Intelligent Physical Agents – see www.fipa.org

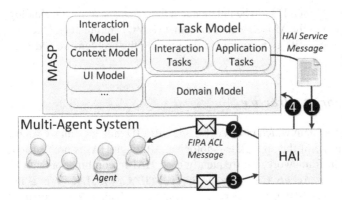

Fig. 2 Architecture of the system, consisting of three top-level tiers: MASP, HAI and the MAS.

and responds by a message as well (3). As a result, HAI notifies the interpreter that the task has been accomplished and updates the data model if necessary (4). Finally, the runtime system of MASP forces the view to apply to the potential changes.

4 Proof of Concept

As we strongly believe, that a combination of MBUID and AOSE facilitates the development of AAL environments, we proceed by describing an agent-based AAL service which we developed within the *SmartSenior* project[5]. As security is an inherent part of the AAL vision we recently implemented an agent-based assistant which is able to detect anomalies in an AAL environment and to inform a user about them by different interaction modalities [12]. The assistant comprises three logical components: sensor-, analysis- and reaction unit. While the sensor unit for itself is not agent-based, the others are. The analysis unit is a multi-agent system consisting of a situation recognition agent, that collects sensor data, a rule evaluation agent and an anomaly detection agent to manage situations which are not covered by rules. While the rule evaluation agent reacts on same event types always in the same fashion, the anomaly detection agent is able to learn a user's regular behaviour and reacts when discrepancies to this regular behaviour are identified using statistical methods. Once an event occurs, it is forwarded to the reaction component. The reaction component is an agent that determines the best interaction possibility to inform the user. This process is straightforward and makes only use of the localisation data gathered from the sensor unit. MASP was utilised to develop and furnish three different UIs [17]. Once the user is close to a screen, a popup will provide informations and options about an occurring event. In case the user is not in scope of a screen but at home, a bracelet the user wears informs about the event. Finally,

---

[5] SmartSenior – longer independence for senior citizens, see www.smart-senior.de

in case the user is currently not at home, the users smart phone is used as interaction device.

5   Conclusion

In this paper, we introduced an approach that facilitates the development of device- and modality independent UIs for agent applications. In order to do this, we argued that MBUID is a suitable foundation for the implementation of agent applications for AAL environments. Furthermore, we emphasised that this area of research is only sparely covered and that existing solutions have severe shortcomings, on either the agent- or on the MBUI side. We further described how the task-model, that is available in most MBUID environments, can be utilised to apply model driven techniques for the development of UIs for agents. Subsequently, we provided an outline of our example system, that makes use of MASP as MBUID environment and JIAC as agent framework. In order to exploit the MBUIs of the MASP for JIAC agent systems, we made use of HAI. Due to HAIs independence from specific UI technologies or agent frameworks, our approach is easily adaptable to other MBUID environments and agent frameworks. After presenting our approach, we illustrated a proof-of-concept implementation of an agent-based AAL service.

It is our opinion that, as sophisticated model driven UI techniques are, the support for the dynamics and the capabilities of distributed multi-agent systems for AAL environments often fall short. To counter this issue, we extended the CTT in a way that allows agents to act proactively. We also used the MVC paradigm to strictly separate between agent and UI specific parts. Nevertheless, currently there are unsolved issues regarding our approach. While agent technology offers capabilities for coordinated and orchestrated the increasing number of applications in AAL environments, a mechanism is required to ensure a reasonable UI. Hence, a first step is to find a more sophisticated possibility to negotiate about the best way of interaction for a given situation, even for the whole AAL environment.

Evaluating design-oriented approaches is usually a tedious task. In the future, we want to demonstrate that a combination of MBUID and AOSE facilitates the development of AAL environments. In order to measure the impact on the development effort of AAL services, we intent to compare different developer teams with equal capabilities performing the same task – some using the presented approach, the others not. We will present the results of this evaluation in a succeeding work.

References

1. Agent Oriented Software Pty. Ltd.: JACK Intelligent Agents – WebBot Manual, 5.3 edn. Agent Oriented Software Pty. Ltd., Victoria, Australia (2009)
2. Ahrndt, S., Lützenberger, M., Heßler, A., Albayrak, S.: HAI – A Human Agent Interface for JIAC. In: Klügl, F., Ossowski, S. (eds.) MATES 2011. LNCS, vol. 6973, pp. 149–156. Springer, Heidelberg (2011)

3. Blumendorf, M., Feuerstack, S., Albayrak, S.: Multimodal user interfaces for smart environments: The multi-access service platform. In: Bottoni, P., Levialdi, S. (eds.) Proceedings of the Working Conference on Advanced Visual Interfaces. ACM (2008)

4. Braubach, L., Pokahr, A., Moldt, D., Bartelt, A., Lamersdorf, W.: Tool-supported interpreter-based user interface architecture for ubiquitous computing. In: P. Forbrig, Q. Limbourg, B. Urban, J. Vanderdonckt (eds.) Interactive Systems - Design, Specification, and Verification, pp. 89–103. Springer (2002)

5. Calvary, G., Coutaz, J., Thevenin, D., Limbourg, Q., Bouillon, L., Vanderdonckt, J.: A unifying reference framework for multi-target user interfaces. Interacting with Computers 15(3), 289–308 (2003)

6. Corchado, J.M., Perez, J.B., Hallenborg, K., Golinska, P., Corchuelo, R. (eds.): Workshop on Agents for Ambient Assisted Living. AISC, vol. 90. Springer, Heidelberg (2011)

7. Dutton, W.H., Blank, G.: Next generation users: The internet in britain. Oxford Internet Institute, University of Oxford (2011)

8. Eisenstein, J., Rich, C.: Agents and guis from task models. In: Proceedings of the 7th International Conference on Intelligent User Interfaces, pp. 47–54. ACM (2002)

9. Feuerstack, S., Blumendorf, M., Albayrak, S.: Prototyping of Multimodal Interactions for Smart Environments Based on Task Models. In: Mühlhäuser, M., Ferscha, A., Aitenbichler, E. (eds.) AmI 2007 Workshops, CCIS, vol. 11, pp. 139–146. Springer, Heidelberg (2008)

10. Hirsch, B., Konnerth, T., Heßler, A.: Merging agents and services – the JIAC agent platform. In: Bordini, R.H., Dastani, M., Dix, J., Amal, E.F.S. (eds.) Multi-Agent Programming: Languages, Tools and Applications, pp. 159–185. Springer, Heidelberg (2009)

11. Limbourg, Q., Vanderdonckt, J., Michotte, B., Bouillon, L., López-Jaquero, V.: USIXML: A Language Supporting Multi-path Development of User Interfaces. In: Bastide, R., Palanque, P., Roth, J. (eds.) DSV-IS 2004 and EHCI 2004. LNCS, vol. 3425, pp. 200–220. Springer, Heidelberg (2005)

12. Mustafić, T., Clausen, J., Messerman, A., Chinnow, J.: Concept of a sensor based emergency detection in a home environment. In: Ambient Assisted Living 2010, p. 4. VDE Verlag (2010)

13. Paterno, F., Mancini, C., Meniconi, S.: Concurtasktrees: A diagrammatic notation for specifying task models. In: Howard, S., Hammond, J., Lindgaard, G. (eds.) Proceedings of Interact 1997. Human-Computer Interaction Conference. Chapman and Hall (1997)

14. Paterno, F., Santoro, C., Spano, L.D.: MARIA: A universal, declarative, multiple abstraction-level language for service-oriented applications in ubiquitous environments. ACM Transactions on Computer-Human Interaction (TOCHI) 16(4), 1–30 (2009)

15. Pokahr, A., Braubach, L.: The Webbridge Framework for Building Web-Based Agent Applications. In: Dastani, M., El Fallah Seghrouchni, A., Leite, J., Torroni, P. (eds.) LADS 2007. LNCS (LNAI), vol. 5118, pp. 173–190. Springer, Heidelberg (2008)

16. Pruvost, G., Bellik, Y.: Ambient multimodal human-computer interaction. In: Proceedings of the Poster Session at The European Future Technologies Conference, pp. 1–2 (2009)

17. Raddatz, K., Schmidt, A.D., Thiele, A., Chinnow, J., Grunnewald, D., Albayrak, S.: Sensor-based detection and reaction in ambient environments. In: Ambient Assisted Living 2012. VDE Verlag (to appear, 2012)

18. Tran, V., Kolp, M., Vanderdonckt, J., Wautelet, Y., Faulkner, S.: Agent-Based User Interface Generation from Combined Task, Context and Domain Models. In: England, D., Palanque, P., Vanderdonckt, J., Wild, P.J. (eds.) TAMODIA 2009. LNCS, vol. 5963, pp. 146–161. Springer, Heidelberg (2010)

Virtual Agents in Next Generation Interactive Homes

Rafael Del-Hoyo, Luis Sanagustín, Carolina Benito, Isabelle Hupont,
and David Abadía

Abstract. Today's houses are slowly turning into a complex electronic net of
devices. The increasing complexity of systems and the need for these systems to
remain simple, accessible and transparent for the user, makes it necessary to
research technologies that enable intelligent and autonomous computing and new
ways of interacting with future home. Autonomic computing systems are those
which can manage themselves given high level objectives. If we integrate
autonomic computing and new interactive user mechanisms like virtual agents, we
obtain the future smart homes.

Keywords: Autonomic Computing, HCI, Virtual Agents, Smart Homes.

1 Introduction

Nowadays, huge R&D efforts are running on the re-invention of the Internet so
that it is able to cope with future challenges, like the viral growth of the number of
connected users, devices, services and user-generated contents. Today's houses
are slowly turning into a complex electronic net of devices. Multimedia TVs based
in DLNA (Digital Living Network Alliance), sensors, automation controls and
energy consumption meters are connected to the Internet via residential gateways,
using a variety of home communication networks (fiber, WiFI, Power Line). As a
resultof this people is going to beimmersed in the Internet of things paradigm or
the Internet of objects connected inside of each home. The increasing complexity
of systems and the need for these systems to remain simple, accessible and
transparent for the user, makes it necessary to research technologies that enable
intelligent and autonomous computing and new ways of interacting with future

Rafael Del-Hoyo · Luis Sanagustín · Carolina Benito · Isabelle Hupont · David Abadía
Instituto Tecnológico de Aragón, P.T. Walqa Ctra. Zaragoza, N-330a, Km 566,
Cuarte (Huesca), Spain
e-mail: {rdelhoyo,lsanagustin,cbenito,ihupont,dabadia}@ita.es

J.M.C. Rodríguez et al. (Eds.): Trends in PAAMS, AISC 157, pp. 9–17.
springerlink.com   © Springer-Verlag Berlin Heidelberg 2012

home. Autonomic computing systems are those which can manage themselves given high level objectives. These systems include environments that are able to evolve without the need for human interaction. These environments are capable of installing, configuring, maintaining and healing themselves, and their own components. This paper presents an Autonomic Interactive Fusion engine platform developed in the GENIO[1] project. The main element of the architecture is the Intelligent Autonomous System in charge of the information fusion process. This also controls a virtual agent that allows increases the interactivity from user perspective. This paper is structured in the following way; the first section is a brief statement about autonomic computing and virtual agent in smart homes. The second section describes current projects in smart homes and GENIO project and the following section describes the intelligent framework used for smart homes. The fourth section explains how the information is joined inside the fusion information engine. Finally, several conclusions are set forth.

1.1 Autonomic Computing

The essence of autonomic computing systems is self-management, or, the ability to reduce human interaction in administration tasks to the minimum. As it is explained in previous research [1], these systems should provide self-configuration, self-optimization, self-healing, self-protection. The system should incorporate itself seamlessly, and the other components present in the system must adapt to its presence by learning new configurations or topologies. An automatic system should continually seek ways to tweak parameters, and, at the same time, should be able to find and apply the lastest updates for each system component. Autonomic systems should detect, trace, diagnose and repair bugs and failures. Autonomic systems should defend themselves from large scale problems arising from malicious attacks or big failures.

1.2 Human Interaction through Virtual Agents

Virtual agents have proved to be a useful way of HCI (Human Computer Interaction). For humans, it is easier to communicate with a computer through a conversation with a virtual agent as opposed to just a keyboard and mouse. To make a virtual agent interact in a consistent, emotionally empathic and intelligent way with the user, a strategy must be defined for recognizing, integrating and interpreting user information coming from different modalities (video, audio, etc.). Secondly, it is important to realize how the human mind works to correctly "model" the virtual agent's reasoning mechanisms. The human brain is characterized by its capacity to handle and store uncertain and confusing perceptions. People usually face problems with great uncertainty and partial, context-dependent, and contradictory information. SoftComputing techniques, in special Fuzzy Logic, make it possible to model these types of problems and to find solutions similar to the ones taken by human beings. In doing so, it is possible to develop a more "cognitive" computation that tackles effectively the interaction among persons and virtual agents, how they communicate and act through words and perceptions [2]. Finally, the virtual agent must be believable: it has to move properly, paying special attention to its facial expressions [3], and have the capacity to talk in natural language [4]. Emotions have been proved to play an essential role in decision making, perception, learning and more [5]. Consequently, besides its external appearance, the virtual agent must possess some affectivity, an innate characteristic in humans, for which it is necessary to carefully manage the emotional display of the virtual agent. Human Computer Interaction (HCI) gets more natural when using a virtual agent as computer side communication entity.

1.3 Smart Autonomous Media Homes State of the Art

One of the most important fields to apply Autonomic Computing and Human Computer Interaction technologies is houses, thus making them intelligent or smart houses. These houses would detect the people inside, self-configure by personalizing the services for each users and detecting and configuring new devices plugged into the house; would self-optimize by disconnecting lights or closing doors if people aren't present; would self-heal by controlling sensors and preventing problems related to physical and software elements; and would self-protect by identifying the current users at home, and preventing external attacks. There are several research efforts to introduce major autonomic capabilities in smart houses, like self-configuration using adaptation models [6], petri-nets [7] or variability models [8] which determine a set of policies to know how the system should be modified against changes at home; or self-protection. The University of Valencia has developed a Model-Based Reconfiguration Engine (MoRE) [8,9]. Finally, the University of Mondragon, the University of Ulster and Washington State University present their project CASAS. It is an adaptive smart home autonomous system that utilizes machine learning and data mining [10] techniques to discover patterns in resident's daily and repetitive activities [11] and to generate automation polices that mimic these patterns. The autonomous system can be guided by resident providing explicit feedback or it can be left to the system to automatically discover and adapt to changes in pattern of activities.

1.4 The Smart Home in GENIO Project

The research presented in this paper is part of the European GENIO project. The GENIO project aims to define the home network of the future, developing an advanced self-management of the home network, facing the problem of heterogeneity and transparency of the devices and their interactions, and looking to maximize automation and to respond intelligently to events and alarms. The project intends also to solve other challenges like ubiquitous access to the home network contents and personalization of the services. A diagram of GENIO smart home can be seen in figure 1.

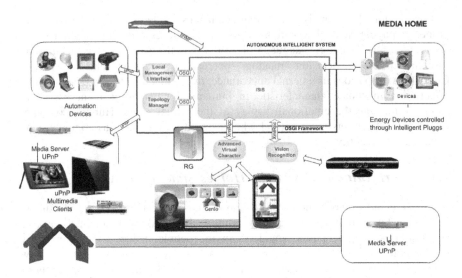

Fig. 1 Diagram of the system working as Autonomous System inside Smart Home.

2 Framework Technology Overview

The aim of this section is to establish the Intelligent Autonomous System's technology overview. One of the possible technologies which is candidate for handling this type of smart home is algorithm based on soft-computing/computational intelligence techniques. These algorithms are able to work with a great number of data (even noisy and incomplete), and they also allow predicting the behavior of highly nonlinear systems, as is the case of Home Systems and in special communication systems. As we have showed in GENIO project, these properties allow us to analyze, predict the state of an IP network or to make decision about any problem inside of the home. In this project, the development of an intelligent decision support system (iDSS) is proposed, called Intelligent Autonomous System (IAS), based on some well-known artificial neural models. IAS will allow the entire home status management, dealing with high dimensionality information, through a new advanced interface personalized into a Virtual Agent. The objective of this virtual agent is to be the human-interface between the home user and the autonomous systems, informing the user about the status of the home and the predicted situations found. Also, through this interface the user can manage the home devices using several other interfaces, like SNMP or uPNP. The control logic in the IAS platform is implemented inside of the intelligent framework using natural language rules (fuzzy rules) and neural network for pattern recognition. This framework is also responsible of the behavior of the virtual agent. This IAS platform is able to evolve and adapt according to the actions obtained from the user. During the learning process, user patterns like the number of user repetitions, watching movies or managing home devices are learned by means of a neural network supervised learning process.

Due to the rules-based intelligent framework, the proposed platform is a powerful tool for general autonomous home systems. The main AI (Artificial intelligence) technologies uses in the decision autonomous system for GENIO are neural networks, Rule Engines and finally AIML(Artificial Intelligence Markup Language).

2.1 The Intelligent Support Interaction System (ISIS)

ISIS is the main element of the proposed system. It is the evolution of a SoftComputing-based intelligent system called PROPHET that enables real-time automatic fuzzy decision making and self-learning over any kind of incoming inputs (from sensors, video channels, audio channels, probes…)[12].ISIS is the engine in charge of the logic of the platform from IAS point of view. It is also the inference engine that makes the virtual Agent to react to different inputs coming to the platform. According to different inputs of the platform, ISIS extracts knowledge and thanks to the use of Neuro-Fuzzy techniques, the module has the capability of self-extracting and self-learning new fuzzy decision rules from historical data. ISIS consists of a set of modules for pre-processing, integrating and extracting information and making decisions in a flexible way under uncertain contexts. It is described in the following modules:

- **Hybrid rule inference engine:** it is the main sub-system of the Autonomous System. It is in charge of rule-based decision-making tasks..
- **Knowledge and data persistence module:** system that manages data and knowledge (rules) information.
- **Integration and transformation module:** module in charge of filtering, synchronizing and pre-processing the incoming inputs, in order to make them compatible with the hybrid rule inference engine input format.
- **Application control module:** state machine that controls the Autonomous Intelligent System behavior.
- **AIML module:** the AIML module computes an appropriate natural language answer, for a given user interaction context.
- **Communication interface with the GUI interface:** that manages communication between the Autonomous System and the GUI.
- **Communication interface with the Topology Managercomponent:** that informs the Autonomous System in real-time about the events that occur in the home network

2.2 Neural Network Algorithms Applied in Our Work

Learning from examples (or historical data) is one of the capabilities that make artificial neural networks and neuro-fuzzy systems a suitable approach for pattern recognition in home environments.We have applied neural networks and neuro-fuzzy to home management and pattern recognition, and we have compared results from both models and we obtain similar results due to the simplicity of the uses cases generated.

2.3 Rule Engine

The embedded rule inference engine is in charge of rule-based decision making tasks in the home manage process. It is a hybrid rule inference engine since it can both deal with crisp rules (applied to exact inputs' values) and execute inference from rules that handle fuzzy concepts. Fuzzy logic is a form of multi-valued logic derived from fuzzy set theory to deal with reasoning that is robust and approximate rather than brittle and exact. Furthermore, when linguistic variables are used, these degrees may be managed by specific functions. The elements in the inference engine's Working Memory are not only the rules pre-defined by an expert, but also the set of automatically self-learned decision rules created by the knowledge extraction and classification obtained from neuro-fuzzy systems. Fuzzy set theory defines fuzzy operators on fuzzy sets. The problem in applying this is that the appropriate fuzzy operator may not be known. For this reason, fuzzy logic usually uses IF-THEN rules. Rules are usually expressed in the form:

```
IF variable IS property THEN action
```

In the case of GENIO project will have the following rules examples

```
IF temperature IS very cold THEN select movie about
winter (personalization rule)
IF QoS is low  THEN don't allow remote movies
IF QoS is high THEN select HD movies
```

There is no "ELSE" – all of the rules are evaluated, because the temperature might be "cold" and "normal" at the same time to different degrees.

3 Multimodal Fusion Engine

The main feature of the Autonomous System is its capability to integrate information coming from several sources, of very different natures and arriving with different timings. The system integrates data (every 500ms) from the following sources:

- **Pattern Recognition:** through the inclusion of different classification algorithms previously predefined. This preprocess module allows the generation of new attributes based on this data mining algorithm and the generation of an attribute that represents the output of this classifier (figure 2).
- **Virtual Agent Speech Recognition:** based on the recognition of the user's voice, it is converted into text and placed as an attribute of the text itself.
- **Presence recognition and number of people counting:** the number of persons found in the scene is introduced as a new attribute.
- **Gesture Recognition:** gesture recognition is performed by a gesture recognition algorithm within the platform Kinect.
- **Feature Image Classification Algorithm:** With a vision algorithm based on a feature detection algorithm called Surft. This information is entered into the system as a new nominal (object name recognized) attribute.

- **Information about UPnP devices / network existing inside the home:** when an uPnP device is detected, ISIS dynamically generates a number of attributes into the system corresponding to each of the properties associated to UPnP device.
- **SNMP information from network devices Home:** ISIS is able to monitor any home network device.
- **Energy consumption Information and home automation sensors:** home sensors are mapped to numeric attributes inside the inference engine.

3.1 Heterogeneous Output Information

The most important actions that can be generated in a given time *t*in the system are:

- **Gestural animation of the virtual actor's face:** with this types of consequences in the rules can be modified emotion the face of virtual agent.
- **VirtualAgentverbalinformation:** the inferenceenginecan control a Text to Speech virtual agent system.
- **UPnP/DLNA device control:** The enginecan operate asUPnP/DLNA controlpointand controlany deviceof this kind.
- **Control home automation devices and Energy:** The element of decision allows you to control alarms or interrupting powerto any electrical appliance.
- **Generate the training of a given data mining classifier:** Based on information from the historic the system launches a learning process for a particular pattern using one specific classifier such as a neural network.

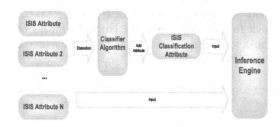

Fig. 2 Integration of the pattern recognition algorithms inside the inference engine

Fig. 3 Examples of Agent's Emotional expressions.

3.2 New User Home Interaction

The user interaction as previously discussed is performed by integrating information from heterogeneous data sources. The user can send audio (voice commands) or visual information (detection people, generate hand gestures or detect objects that are recognized by the system). This information is processed and integrated with other types of information to generate verbal responses or emotions by the virtual agent to increase interaction between man and machine. Different use cases have been tested. From the generation of alarms by the virtual agent through the recognition of patterns in sensors deployed around the house or the management of uPnP devices such as movies or cameras via voice commands or through hand gestures. Others more complex cases can be when the user is watching a film on a TV in the lounge, the user shows a Tablet PC to the virtual agent and the film is moved to Tablet at the point where you were watching it and thus bring in your media information always with you and anywhere in the house. Moreover, the network is monitored all time, so you can validate that there is sufficient quality of service on the network before launching the film in the tablet itself. One of the most interesting system features is the ability to generate emotions in the virtual agent depending on the condition of your home. Using Fuzzy rules ISIS can generate a temporary emotional personality of the agent, using a similar method than in [13] therefore different emotional expressions (figure 3) are generated on the agent during user interaction.

4 Conclusions

A novel platform for creating smart interactive homes has been presented. The platform has been achieved as a result of the work developed in GENIO project. Moreover, fuzzy rules used inside the inference engine enable the Home self-optimization and self-healing. Finally the multimodal integration and information fusion and the pattern recognition features allow the Home self-protection and self-healing for diagnosis. The second outstanding feature of GENIO project is the great possibilities it offers to the user for interacting with the Home by means of the Virtual Agent. The agent is able to interact with physical elements of the home, such as sensors or multimedia DLNA devices.

Acknowledgments. This work has been co-funded by GENIO project (TSI-020400-2010-98), QUEEN project (IPT-2011-1235-430000).the Spanish Ministry of Industry, project FEDER ATIC.

References

1. Kephart, J.O., Chess, D.M.: The Vision of Autonomic Computing. Computer, 41–53 (January 2003)
2. Zadeh, L.A.: Computing with Words and Perceptions - A Paradigm. In: Proceedings of the 2003 International Conference on Machine Learning and Applications, pp. 3–5 (2003)
3. Pantic, M., Bartlett, M.S.: Machine Analysis of Facial Expressions. Face Recognition, 377–416 (2007)

4. Cowie, R., Douglas-Cowie, E., Schroeder, M.: Speech and Emotion. A conceptual framework for research. In: Proceedings of the ISCA Workshop, Belfast (2000)
5. Egges, A., Kshirsagar, S., Magnenat-Thalmann, N.: Generic Personality and Emotion Simulation for Conversational Agents. Computer Animation and Virtual Worlds 15, 1–13 (2004)
6. Morin, B., Fleurey, F., Bencomo, N., Jézéquel, J.-M., Solberg, A., Dehlen, V., Blair, G.S.: An Aspect-Oriented and Model-Driven Approach for Managing Dynamic Variability. In: Czarnecki, K., Ober, I., Bruel, J.-M., Uhl, A., Völter, M. (eds.) MODELS 2008. LNCS, vol. 5301, pp. 782–796. Springer, Heidelberg (2008)
7. Zhang, J., Cheng, B.H.C.: Model-Based Development of Dynamically Adaptive Software. In: Proc. 28th Int'l Conf. Software Eng. (ICSE 2006), pp. 371–380. ACM Press (2006)
8. Coplien, J., Hoffman, D., Weiss, D.: Commonality and Variability in Software Engineering. IEEE Software, 37–45 (1998)
9. Cetina, C., et al.: Autonomic Computing Through Reuse Of Variability Models At Runtime: The Case Of Smart Homes. Computer, 37–43 (2009)
10. Cook, D., Holder, L.: Sensor Selection to Support Practical Use of Health-Monitoring Smart Environments. Data Mining and Knowledge Discovery 1(4) (2011)
11. Rashidi, P., Cook, D.: Keeping the resident in the loop: Adapting the smart home to the user. IEEE Transactions on Systems, Man, and Cybernetics, Part A: Systems and Humans 39(5), 949–959 (2009)
12. Martínez, H., Del-Hoyo, R., Sanagustín, L., Hupont, I., Abadía, D., Sagüés, C.: Augmented Reality based Intelligent Interactive e-Learning Platform. In: ICAART (1), pp. 343–348 (2011)
13. Liu, Z.: A Personality Based Emotion Model for Intelligent Virtual Agents. In: 2008 Fourth International Conference on Natural Computation, Jinan, Shandong, China, pp. 13–16 (2008)

Accessing Cloud Services through BDI Agents

Case Study: An Agent-Based Personal Trainer to COPD Patients

Kasper Hallenborg, Pedro Valente, and Yves Demazeau

Abstract. Cloud computing is envisioned a dominant role in the future. Extensive amount of data are stored, applications running in the cloud, and globally accessible. However, users neither cannot nor are interested in observing and processing those amounts of information. Thus, a mentalistic model that virtually represents the user's goals could be integrated with the cloud to present processed extracts in a cognitively accessible way. Such an approach is presented with deliberative BDI agents both in general and in a case study for COPD patients.

1 Introduction

Cloud computing is envisioned to play a dominant role in not only storage of information, but also how we do "computing in the cloud". Still without a clear and unique definition – cloud computing is an overly used concept that evolves along the growing attention to virtualized networked services, Software as a Service (SaaS). Accessibility, availability, and interoperability are central aspects of importance to both fulfill the requirements of users' needs and the high level of machine-machine cooperation.

Personal/Electronic Health Records (PHR/EHR) is a perfect example for the use of cloud computing, where the multi-user / multi-role perspective, privacy, security, and accessibility issues are addressed, as both health-care professionals (e.g. hospitals, practitioners, physicians, etc.) and you personally are involved.

Kasper Hallenborg · Pedro Valente
Maersk McKinney Moller Institute, University of Southern Denmark, 5230 Odense M, Denmark
e-mail: {hallenborg,prnv}@mmmi.sdu.dk

Yves Demazeau
Laboratoired'Informatique de Gre-noble - CNRS, Grenoble, France
e-mail: yves.demazeau@imag.fr

J.M.C. Rodríguez et al. (Eds.): Trends in PAAMS, AISC 157, pp. 19–28.
springerlink.com   Springer-Verlag Berlin Heidelberg 2012

Currently, large initiatives try to settle, with Google Health, Microsoft HealthVault, and Dossia being the most promising clouds that structure PHRs based in standardized medical formats, such as HL7, CCR and in compliances with Continua Health Alliance.

Thus, the ultimate vision of clouds, like Google Health, is a set of interoperable services and cloud applications that brings together health-related information from hospitals, care-givers, etc. with personal recordings and data-gathering sensors into a complex, but coherent health information system.

SaaS spawned from service-oriented architectures and se-mantic web services, which do take advantage of prominent technologies to abstract and increase flexibility of services for the end-user, such as WSDL and OWL ontology. However, the composition of services must still rise from a set of basic action by the user [4]. In particular for PHRs the heterogeneous set of users and the complexity of stored information call for transformations that can present medical information in a cognitively accessible way [5], but analogies can in general be foreseen in other domains as well.

We will propose a deliberative agent model based on the BDI architecture to close the gap between cloud-based services (SaaS) and ubiquitous intelligent services to the user in general. Initially, augmenting the need for deliberative agents over standard service models in respect to a goal-oriented approach of user needs. The proposed model will be applied on a case study of a motivating UI for COPD patients that should perform physical exercises to not worsen their chronic disease.

2 Agents to the Cloud

2.1 Limitations of a Service Based Model

Traditionally, web services were defined as software systems designed to support interoperable machine-to-machine interaction over a network [6]. Even with semantic descriptions of the services (OWL-S) where all dependable service should be known at design time, a service based model does not overcome the limitations in how users naturally will express the goals they want to achieve rather than the actions they wish to be performed [4].

Thus, several key aspects are important to address if user-centered cloud applications should enable cloud computing to become the 5thutility [7] (after water, electricity, gas, and telephony) as individual, pervasive and intelligent services.

- Locus of control – design of applications must be user-centered and reflect the user's assessment of goals to be achieved. E.g. a health advisor app will continuously update its status/advice according to external percepts rather than just presenting external data.
- Cognition – pervasive applications should not rely on the user to process information in respect to current goals. The apps should be aligned with the true vision of Mark Weiser's calm computing of not drawing user's attention unless necessary [8].

- Adaptation – information resources often reside in many systems (clouds) and needs to be combined either to be relevant or make sense for users. Standards do not prevail across all systems, and never will, so adapting, abstracting, and transforming data is required.
- Context – environments are highly dynamic due to users' mobility and global accessibility. Therefore context of the application is to be observed and taken into account. E.g. considering time and location will be important when trying to encourage user to do physical exercises.
- Learning – it is generally accepted that intelligence involves some kind of learning capability, which clearly is the case, if services are not to be experienced as repetitive annoying reminders. Thus, learning support will significantly enhance user acceptance. In particular for health monitoring and exercise guidance the system must adapt to individual preferences and progress.

In the following section we will advocate for the capabilities of deliberative agents vs. semantic web services to cope with the aspects presented.

2.2 Capabilities of Deliberative Agents

The procedural approach of semantic web services built on the basis of synchronous remote procedure calls has been preferred by industry over more abstract definitions and less strict approaches for agent technologies, which comes from asynchronous message passing architectures [4]. However, the goal-oriented agents have a lot more to offer than services founded in basic actions.

From a modeling perspective the autonomous characteristic of agents and the architecture for deliberative agents, e.g. the BDI model, has a more natural mapping to a user-centered design. The BDI model, reflecting the beliefs (current assessment of a world model), desires (objectives to be accomplished) and intentions (current plans being executed), has a very "mentalistic' notion for capturing user's preferences and goals [9], whereas services mostly are designed as information providers for users that needs further (mental) processing. Agents have an internal locus of control, being responsible through its autonomous characteristic to be proactive and react when actions needs to be taken. In contrast objects and services (in its traditional definition) are activated by external invocations. Users should be motivated/guided rather than being instructed.

Goal-directed behaviors are the nature of deliberative agents, and goals can be both fuzzy and abstractly described, such as "user should maintain high level of physical activity", instead of an inflexible specific exercise plan, e.g. "10 minutes of bicycling at level 3 in the morning, walking 1 km afternoon at minimum, …". The difference is huge in respect to user acceptance of the system, as in the latter example the user are guided through basic action elements of the plan or must set contingency plans herself. Whereas the agent approach is not fixed to a specific plan, but all inputs may contribute to the overall goal, and the goal itself is dynamic, e.g. if doctors send updates to the PHR. Thus, there is a need for reasoning on the received percepts and other inputs, if personalized applications should improve of user acceptance. Especially, for healthcare and activity applications users prefer to be in control rather than being instructed.

Interoperability is by definition central to web services, however, despite the profile specification of web services to exchange information in terms of XML formats and initiatives taken on dynamic service discoveries, they are still relying on common understanding and/or adapters for exchanging content information between systems. Even though hard efforts are taken to develop common standards across many areas, a complete and flexible common standard on the content of messages, as rich and powerful as the human interpretation, across all knowledge domains will never exists. However, the agent community is building on knowledge representation and use ontologies to abstractly describe and negotiate on the content of messages, which is a step in the right direction of enriching the possibilities of adaptation across different systems.

Context and context-awareness are the building blocks of pervasive and ubiquitous systems, and slowly progress to have a more dominate role in all ICT system, so services can be experienced as more intelligent. It is clear that reasoning and combining context information will have a great influence on agents capabilities. Recently, the importance and influence of integrating context to the BDI model has been advocated by Koch and Dignum[11].

In many years learning was not really considered of high importance in the agent community, but has gained more attention during the last decade [12]. While e.g. complex neural and Bayesian networks might be the reasoning engine in some cognitive agent systems, a more static knowledge representation might be enough for other systems. However, the advantages of an updated and adapting knowledge space in agent systems are unquestionable.

3 The Agent Model

Deliberative agents extend the basic characteristics of agents given by Jennings and Woolridge (autonomous, reactive, proactive, and social) [13] with knowledge representation and symbolic reasoning capabilities [14]. So while, the locus of control is well captured in a mentalistic mapping based on the autonomous characteristic of agents, the strong reasoning engine of deliberative agents covers the need for cognition, including learning aspects. Adaptation and context-awareness is well aligned with reactive and proactive behaviors of agents and their responsiveness in general.

The BDI model by Rao and Georgeff[15] is by far the most applied architecture for deliberative agents, and it is conceptual simple and appropriate to describe bridging pervasive agents to services in the clouds. Therefore, we will use the BDI model as the base architecture in the following discussion, however the design could with minor effort be transformed to other cognitive models for agents, as the user-centered mentalistic approach is the essence of closing the gap. With the BDI model being the most applied and best supported, this was a natural choice for both the general model and in our case study.

The scope of the paper and the presented model is to cover use of cloud computing, where information is extracted from the clouds (or other information sources) either in a reactive or proactive manor. After some abstraction, transformation, and/or reasoning it is presented to the user. Whether agents exist in the cloud, and thereby respect the definition of SaaS, or runs on e.g. the end-user's mobile devices connected to the cloud is not essential to the approach, but in implementation there will be a difference in how local context information is integrated. In relation to the operational logic for the model presented below, the goals of the agents are then constrained to visualization tasks that aim to motivate the user to address or improve the current observations that defines the current belief set.

The operational logic of BDI agents was formalized by Rao with the Agent-Speak(L) language [16], (earlier approaches exist, e.g. by Cohen & Leveque [17]), which later on have materialized in agent languages such as Jason, 2APL, and Jadex. Variations over the notion exist that include more than the basic predicative of beliefs and goals, so we will partly follow the notion used in [11]. Therefore we can define the BDI agent by the tuple$\langle \mathcal{B}, \mathcal{D}, \mathcal{J}, \Pi \rangle$. Where $\mathcal{B}$ is the belief base, $\mathcal{D}$ is the desire base, $\mathcal{J}$ the intention base, and Π is the set of selection functions and plans, which comes out as the practical rules of the system that can be expressed in the form $\varphi \leftarrow \epsilon | \pi$. φ is the goal, π the concrete plan of action that should lead to the goal, and ϵ the event that triggers the selection function.

The selection functions can be categorized for beliefs, desires, and updates for the intension base due to changes in the belief and desire base caused by the other triggers. However, triggers to update the belief or desire base are usually how the external input comes in. In this case the information is delivered by SaaS of the cloud.

So, whereas the BDI model structure the internal reasoning of the agents and how knowledge about the world is represented, the inter-agent and external communication is related to the social capabilities of the agents, which also include agent-cloud communication. In most agent systems the communication is based on the speech act theory originally defined by Searle [21], and communication protocols have furthermore been standardized by FIPA.

In terms of the agent-cloud communication with respect to the scope for SaaS and agents defined above, the agents would either reactively observe changes in the cloud or proactively request or query the cloud. It corresponds to the **subscribe**, **request**, and **query** protocols of the FIPA standards. It is not within the focus and scope of this paper to explain setup and initialization of binding agents and clouds. Basically, it is done by setting up gateway agents that translate between SOAP messages of web services and ACL messages of agents. WSIG is an example of such a tool to create gateway agents for the JADE agent platform, it can even be used to map WSDL to FIPA interaction protocols.

Agent communication is message-based and uses an agent communication language (ACL). The content of messages can be encoded using ontologies of the domain, so the knowledge sent to the agent can be described in rather abstract terms. Thus, the triggering events for a state change of the agent will either come from inform messages of the FIPA protocols mentioned above, as **subscribe**, **request**, and **query** protocols all use ACL messages with the performative set to inform for the response.

First the inform message will triggers the selection function on the belief base, which formally is given by

$$S_{\mathcal{B}}(\langle \mathcal{B}, \mathcal{D}, \mathcal{I}, \Pi \rangle) = \langle \text{inform}, id, at \rangle$$
$$\overline{\langle \mathcal{B}, \mathcal{D}, \mathcal{I} \rangle \rightarrow \langle \mathcal{B} \cup \{at\}, \mathcal{D}, \mathcal{I} \rangle}$$

Where id is the identifier of the agent and at is the predicates expressed in first order logic, as explained in [16].

In the deliberation engine of the agent it will consider the desire base if there are relevant plans that should be added to the intention base, and such a plan would be considered applicable if a unifier θ exists for the plan, so the plan is a logical consequence of the current belief base with respect to the event. Formally, we can describe it as the following update to the intention base of the agent.

$$\langle \varphi, C \rangle \in \mathcal{D}, \text{impact}(\beta, \varphi, start), \mathcal{B} \models C\theta, \mathcal{I} \not\models \varphi\theta$$
$$\overline{\langle \mathcal{B}, \mathcal{D}, \mathcal{I} \rangle \rightarrow_{\theta} \langle \mathcal{B}, \mathcal{D}, \mathcal{I} \cup \{\langle \varphi\theta, C, \pi \rangle\} \rangle}$$

Where C is the commitment of the belief base, and $\text{impact}(\beta, \varphi, start)$ is the evaluation of a given change in the belief base, β, will lead to that the desire of the plan φ will be followed, as further described in [11]. This is where the evaluation of context information can filter events from change the current intention base, e.g. to suggest activities for the user during the night.

If the evaluations are successful, these actions of the new concrete plan in the intention set, π, may lead changes in the visualization of the current status for the user for the scope of agent activities we are aiming for.

4 A Case Study

Healthcare systems and approaches are quite different around the world, especially between Europe and the US. However, the need to store health related information and PHRs are universal and can to a large extent be considered separate from the organization of the healthcare system. In particular if our focus is on personalized rehabilitation and training in collaboration with healthcare professionals.

Our case study is actually built on the architecture of a healthcare system for the US, which in still not established in Europe due to legislative issue, but similar setups exist. An overview of the system is presented in Fig. 1.

It is commonly known that especially the developed world will be challenged extensively during the next decades due to the demographic development, which basically mean that the share of elders (65+) will double over the next 30 years, and the ratio between the workforce and retired people will change from approx. 4:1 to 2:1. Potentially, costs of healthcare services will explode, as the costs are highly age-related. In particular expenses to chronic diseases, such as chronic obstructive pulmonary disease (COPD/COLD), diabetes, cardiovascular disorder and Alzheimer's, will be exploding if we do not reconsider treatment programs and benefit from ICT. In particular for COPD patients physical exercises are very import to prevent further progression of the disease. According to[24], it was estimated that 2.74 million people died of COPD worldwide in 2000.

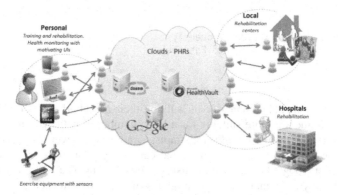

Fig. 1 Overview of healthcare system with PHRs in the clouds

Unfortunately, such training sessions are only provided to a very low number of patients, and often too late in their disease course. However, recent year of advancement in sensor technologies brings forward new opportunities for monitored training at home, and by solving the integration with the PHRs the training and home monitoring can not only save healthcare costs, but also significantly improve capabilities and quality of life for the COPD patients.

4.1 Our Approach

COPD patients are no different than the rest of us – it is very hard to keep up the motivation for physical exercising, unless someone constantly encourages and pushes us to do so (a personal trainer/physiotherapist) or the results/effect becomes very visual. For COPD patients another factor – the anxiety caused by breathing troubles – leads to worse submaximal exercising. Thus, we need a responsive and proactive feedback that can motivate to keep up the training.

Fortunately, recent years advancement of sensors and wireless technologies are currently enabling the vision of home monitoring, and equally important is the integration with open cloud-based PHR systems. Bluetooth based blood pressure meters, weight scales, pill boxes, pedometers, etc. are just some examples of technologies available and already prepared to stream data directly into your health record in Google Health on the user side. Hospitals, pharmacies, etc. can also stream data and medical records into Google Health.

Integration with Google Health is rather simple, as it is part of the Google Data Protocol, which support access to the cloud using standardized cloud protocols, such as Atom Publishing Protocol (AtomPub) or JavaScript Object Notation (JSON). Using the **HealthService** and **HealthQuery** classes of the Google Data API access to your health can be established and data extracted, e.g. using the C# client library:

```
HealthServicehService = newHealthService("appname");
hService.setUserCredentials("username", "password");
HealthQuery query = newHealthQuery();
```

```
// Setup query

AtomFeed feed = hService.Query(query);
foreach (AtomEntry entry infeed.Entries)
  {
     ...
  }
```

Not all COPD patients are capable of walking their neighborhood, so monitoring exercises at home are more suitable for a large number of patients. Such exercise programs are provided by the physiotherapists, and we are developing a prototype based on the Microsoft Kinect camera that can verify your exercises and stream the data to the Google Health account using another agent. The API for the Kinect camera allows you to analyze the joint angles of your body in each frame, which gives the possibility to map the performance against an exercise program.

It is the responsibility of the BDI COPD trainer agent to aggregate the inputs from Google Health and positively respond to the user with a feedback.

Finally, from the professional healthcare system the medical records and diagnose data of COPD is registered in the personal health records. However, as training programs are not directly supported in the Google Health it is the responsibility of the agent to reason on the extracted information and compare it against the general goals of COPD patients that is represented in the desire base of the agent. In that respect the cloud is a source of information and collection of assessments on health conditions, which emphasize the importance of having deliberative agents that can use those inputs in the reasoning engine and present a motivating feedback to the patients. The approach is highly user-centered and flexible compared to training programs that most likely will not be followed precisely anyway without the supervision of e.g. a physiotherapist. By this approach the agent has the freedom (in the virtualization of the end user) to use alternative combinations of exercises and physical activity to reach the goals.

The COPD trainer agent will update the belief base based on inputs that are received from the cloud according to (1). If the evaluation of the change in the belief base, including current context information, enables a new plan to be followed it will be added to the intention base of the agent according to (2). The applicable plan will contain tasks to update the UI and feedback to the user.

With modifications to the desire base the agents can with minor effects be reused as "back-office" agents for the professional healthcare staff for the long-term assessment of the patients conditions. The agent would also be able to filter out those patients that require further attention.

5 Conclusion

We have presented a BDI-based agent model to close the bridge between cloud computing and user-centered applications that aim to present rich and complex information from clouds in a cognitively accessibly visualization to the end-users. In the approach taken the agents are integrated with the cloud to observe changes in the stored information and subscribe to notices that can influence the current belief base of the agent. The mentalistic BDI model envisions a high level of user acceptance for the visualizations.

The presented model has been used to build a prototype system of an agent-based personal trainer for COPD patients that extract and reason on information from the cloud to give a feedback to the user, so they are motivated to continue their important personal training.

We will continue improving the work on the prototype for the COPD patients and the "back-office" agents to the healthcare professionals. This includes further work on the UI and the motivating feedback, and enrichment of the plan in the desire base, however that is not foreseen to influence the principles of the model of bridging agents and cloud computing.

Acknowledgement. The work presented is supported by the Danish Ministry of Science, Technology, and Innovation and is part of a large national project aiming to develop ICT-based solutions to improve quality-of-life and independent living for chronic patients.

References

[1] Dickinson, I., Wooldridge, M.: Agents are not (just) web services: considering BDI agents and web services. In: Proc. of the 2005 Workshop on Service-Oriented Computing and Agent-Based Engineering (SOCABE 2005), Utrecht, The Netherlands (2005)

[2] Sunyaev, A., Chornyi, D., Mauro, C., Krcmar, H.: Evaluation Framework for Personal Health Records: Microsoft HealthVault vs. Google Health. In: Proc. of 43rd Hawaii International Conference on System Sciences, Koloa, Kauai, Hawaii (2010)

[3] Booth, D., et al.: Web Services Architecture (February 2004), http://www.w3.org/TR/ws-arch/

[4] Buyya, R., Yeo, C.S., Venugopal, S., Broberg, J., Brandic, I.: Cloud computing and emerging IT platforms: Vision, hype, and reality for delivering computing as the 5th utility. Future Generation Computer Systems 25(6), 599–616 (2009)

[5] Weiser, M., Brown, J.S.: Designing Calm Technology (December 1995), http://www.ubiq.com/hypertext/weiser/calmtech/calmtech.html

[6] Costantini, S.: Agents and Web Services. The ALP Newsletter 21(2-3) (August 2008)

[7] Koch, F., Dignum, F.: Enhanced Deliberation in BDI-Modelled Agents. In: Demazeau, Y., Dignum, F., Corchado, J.M., Pérez, J.B. (eds.) Advances in PAAMS. AISC, vol. 70, pp. 59–68. Springer, Heidelberg (2010)

[8] Alonso, E., d'Inverno, M., Kudenko, D., Luck, M., Noble, J.: Learning in Multi-Agent Systems. The Knowledge Engineering Review 16(3) (September 2001)

[9] Wooldridge, M., Jennings, N.: Intelligent Agents: Theory and Practice. Knowledge Engineering Review 10(2), 115–152 (1995)

[10] Wooldridge, M.: Reasoning about Rational Agents. MIT Press (2000)

[11] Rao, A.S., Georgeff, M.P.: BDI Agents: From Theory to Practice. In: Proceedings of the First International Conference on Multi-agent Systems (ICMAS 1995), pp. 312–319 (1995)

[12] Rao, A.S.: BDI Agents Speak Out in a Logical Computable Language. In: Perram, J., Van de Velde, W. (eds.) MAAMAW 1996. LNCS, vol. 1038, pp. 42–55. Springer, Heidelberg (1996)

[13] Cohen, P.R., Levesque, H.J.: Intention Is Choice with Commitment. Artificial Intelligence 42(3), 213–261 (1990)

[14] Searle, J.R.: Speech Acts. Cambridge University Press, Cambridge (1969)

[15] COPD International (2004), http://www.copd-international.com

Evaluating the n-Core Polaris Real-Time Locating System in an Indoor Environment

Dante I. Tapia, Óscar García, Ricardo S. Alonso, Fabio Guevara, Jorge Catalina, Raúl A. Bravo, and Juan Manuel Corchado

Abstract. Context-aware technologies allow Ambient Assisted Living developments to automatically obtain information from users and their environment in a distributed and ubiquitous way. One of the most important technologies used to provide context-awareness is Wireless Sensor Networks (WSN). Wireless Sensor Networks comprise an ideal technology to develop Real-Time Locating Systems (RTLS) aimed at indoor environments, where existing global navigation satellite systems do not work correctly. In this regard, n-Core Polaris is an indoor and outdoor RTLS based on ZigBee WSNs and an innovative set of locating and automation engines. This paper presents the main components of the n-Core Polaris, as well as some experiments made in a real scenario whose results demonstrate the effectiveness of the system in indoor environments.

Keywords: Ambient Assisted Living, Real-Time Locating Systems, Wireless Sensor Networks, ZigBee, Web Services.

1 Introduction

Ambient Assisted Living (AAL) tries to adapt the technology to the people's needs by means of omnipresent computing elements which communicate among

Dante I. Tapia · Óscar García · Ricardo S. Alonso · Fabio Guevara · Jorge Catalina
R&D Department, Nebusens, S.L., Scientific Park of the University of Salamanca, Edificio M2, Calle Adaja, s/n, 37185, Villamayor de la Armuña, Salamanca, Spain
e-mail: {dante.tapia,oscar.garcia,ricardo.alonso,
    fabio.guevara,jorge.catalina}@nebusens.com

Raúl A. Bravo · Juan M. Corchado
Department of Computer Science and Automation, University of Salamanca, Plaza de la Merced, s/n, 37008, Salamanca, Spain
e-mail: {raulabel,corchado}@usal.es

J.M.C. Rodríguez et al. (Eds.): Trends in PAAMS, AISC 157, pp. 29–37.
springerlink.com   © Springer-Verlag Berlin Heidelberg 2012

them in a ubiquitous way [1]. In this sense, the continuous advancement in mobile computing makes it possible to obtain information about the context and also to react physically to it in more innovative ways.

Wireless Sensor Networks (WSN) are used for gathering the information needed by AAL environments, whether in home automation, industrial applications or smart hospitals. One of the most interesting applications for WSNs is Real-Time Locating Systems (RTLS). The most important factors in the locating process are the kinds of sensors used and the techniques applied for the calculation of the position based on the information recovered by these sensors. Although outdoor locating is well covered by systems such as the current GPS (Global Positioning System) or the future Galileo, indoor locating needs still more development, especially with respect to accuracy and low-cost and efficient infrastructures [2]. Therefore, it is necessary to develop Real-Time Locating Systems that allow performing efficient indoor locating in terms of precision and optimization of resources. This optimization of resources includes the reduction of the costs and size of the sensor infrastructure involved on the locating system. In this sense, the use of optimized locating techniques allows obtaining more accurate locations using even fewer sensors and with less computational requirements [2].

This paper is structured as follows. The next section explains the problem description, as well as the most widely used wireless technologies to build indoor RTLSs. Then, the main characteristics of the innovative n-Core Polaris system are depicted. After that, some experiments carried out in a real scenario to test the performance of different indoor RTLSs are described, as well as the results obtained by n-Core Polaris. Finally, the conclusions obtained so far are depicted.

2 Problem Description

The emergence of Ambient Assisted Living involves substantial changes in the design of systems, since it is necessary to provide features which enable a ubiquitous computing and communication and also an intelligent interaction with users [1]. This kind of interaction is achieved by means of technology that is embedded, non-invasive and transparent for users. In this regard, users' locations given by Real-Time Locating Systems represent key context information to adapt systems to people's needs and preferences.

Real-Time Locating Systems can be categorized by the kind of its wireless sensor infrastructure and by the locating techniques used to calculate the position of the tags (i.e., the locating engine). This way, there is a combination of several wireless technologies, such as RFID, Wi-Fi, UWB and ZigBee, and also a wide range of locating techniques that can be used to determine the position of the tags. Among the most widely used locating techniques we have signpost, fingerprinting, triangulation, trilateration and multilateration [3] [4].

A widespread technology used in Real-Time Locating Systems is Radio Frequency IDentification (RFID) [5]. In this case, the RFID readers act as *exciters* transmitting continuously a radio frequency signal that is collected by the RFID tags, which in turn respond to the readers by sending their identification numbers. In these kinds of locating systems, each reader covers a certain zone through its radio frequency signal, known as *reading field*. When a tag passes through the reading field of the reader, it is said that the tag *is* in that zone.

Locating systems based on Wireless Fidelity (Wi-Fi) take advantage of Wi-Fi WLANs (Wireless Local Area Networks) working in the 2.4GHz and 5.8GHz ISM (Industrial, Scientific and Medical) bands to calculate the positions of the mobile devices (i.e., tags) [6]. A wide range of locating techniques, then, can be used for processing the Wi-Fi signals and determining the position of the tags, including signpost, fingerprinting or trilateration. However, locating systems based on Wi-Fi present some problems such as the interferences with existing data transmissions and the high power consumption by the Wi-Fi tags.

Ultra-Wide Band (UWB) is a technology which has been recently introduced to develop these kinds of systems. As it works at high frequencies (the band covers from 3.1GHz to 10.6 GHz in the USA) [7], it allows to achieve very accurate location estimations. However, at such frequencies the electromagnetic waves suffer a great attenuation by objects (e.g., walls) so its use in indoor RTLS systems presents important problems, especially due to reflection and multipath effects.

ZigBee is another interesting technology to build RTLSs. The ZigBee standard is specially intended to implement Wireless Sensor Networks and, as Wi-Fi, can work in the 2.4GHz ISM band, but also can work on the 868–915MHz band. Different locating techniques based on RSSI and LQI can be used on ZigBee WSNs (e.g., signpost or trilateration). Moreover, it allows building networks or more than 65,000 nodes in star, cluster-tree and mesh topologies [3]. ZigBee is, indeed, the wireless technology selected for our research.

3 The n-Core Polaris Real-Time Locating System

n-Core Polaris is an innovative indoor and outdoor Real-Time Locating System based on the n-Core platform that features an outstanding precision, flexibility and automation integration [8] [9]. The new n-Core Polaris exploits the potential of the n-Core platform, taking advantage of the advanced set of features of the n-Core Sirius devices and the n-Core Application Programming Interface [8].

The wireless infrastructure of n-Core Polaris is made up of several ZigBee nodes (i.e., tags, readers and sensor controllers) called n-Core Sirius A, Sirius B and Sirius D [8]. They all have 2.4GHz and 868/915MHz versions and include a USB port to charge their battery or supply them with power. Likewise, the USB port can be used to update the firmware of the devices and configure their parameters from a computer running a special application intended for it. On the one hand, n-Core Sirius B devices are intended to be used with an internal battery and include two general-purpose buttons. On the other hand, n-Core Sirius D devices are aimed at being used as fixed ZigBee routers using the main power supply through a USB adaptor. In the n-Core Polaris RTLS, n-Core Sirius B devices are used as tags, while n-Core Sirius D devices are used as readers. This way, n-Core Sirius B devices are carried by users and objects to be located, whereas n-Core Sirius D devices are placed at ceilings and walls to detect the tags. Finally, Sirius A devices incorporate several communication ports (GPIO, ADC, I2C and UART through USB or DB-9 RS-232) to connect to distinct devices, including almost every kind of sensor and actuator. All Sirius devices include an 8-bit RISC (Atmel ATmega 1281) microcontroller with 8KB of RAM, 4KB of EEPROM and 128KB of Flash memory and an IEEE 802.15.4/ZigBee transceiver (Atmel AT86RF230).

In Figure 1 it can be seen the basic architecture of the n-Core Polaris Real-Time Locating System. The kernel of the system is a computer that is connected to a ZigBee network formed by n-Core Sirius devices. That is, the computer is connected to an n-Core Sirius D device through its USB port. This device acts as coordinator of the ZigBee network. The computer runs a web server module that makes use of a set of dynamic libraries, known as n-Core API (Application Programming Interface). The API offers the functionalities of the ZigBee network. The web server module offers a set of innovative locating techniques provided by the n-Core API. On the one hand, the computer gathers the detection information sent by the n-Core Sirius D acting as readers to the coordinator node. One the other hand, the computer acts as a web server offering the location info to a wide range of possible client interfaces. In addition, the web server module can access to a remote database to obtain information about the users and register historical data, such as alerts and location tracking.

Fig. 1 The Web Services based architecture of the n-Core Polaris RTLS.

The operation of the system is as follows. Each user or object to be located in the system carries an n-Core Sirius B acting as tag. Each of these tags broadcasts periodically a data frame including, amongst other information, its unique identifier in the system. The rest of the time these devices are in a sleep mode, so that the power consumption is reduced. This way, battery lifetime can reach even several months, regarding the parameters of the system (broadcast period and transmission power). A set of n-Core Sirius D devices is used as readers throughout the environment, being placed on the ceiling and the walls. The broadcast frames sent by each tag are received by the readers that are close to them. This way, readers store in their memory a table with an entry per each detected tag. Each entry contains the identifier of the tag, as well as the RSSI (Received Signal Strength Indication) and the LQI (Link Quality Indicator) gathered from the broadcast frame reception. Periodically, each reader sends this table to the coordinator node connected to the computer. The coordinator forwards each table received from each reader to the computer through the USB port. Using these detection information tables, the n-Core API applies a set of locating techniques to estimate the position of each tag in the environment. These locating techniques include signpost, trilateration, as well as an innovative locating technique based on fuzzy logic.

Then, the web server module offers the location data to remote client interfaces as web services HTTP (Hypertext Transfer Protocol) over SOAP (Simple Object Access Protocol). Figure 2 shows a screenshot of the web client interface. This client interface has been designed to be simple, intuitive and easy-to-use. Through the different interfaces, administrator users can watch the position of all users and objects in the system in real-time. Furthermore, administrators can define restricted areas according to the users' permissions. This way, if some user enters in an area that is forbidden to it regarding its permissions, the system will generate an alert that is shown to the administrator through the client interfaces. In addition, such alerts are registered into the database, so administrators can check anytime if any user violated its permissions. Likewise, administrators can query the database to obtain the location track of a certain user, obtaining statistical measurements about its mobility or the most frequent areas where it moves.

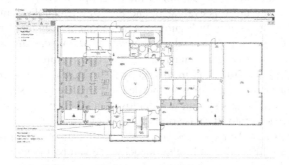

Fig. 2 Web client interface of the n-Core Polaris system.

Furthermore, users can use one of the general-purpose buttons provided by the n-Core Sirius B devices to send an alert to the system. Similarly, administrators can send alerts from the system to a user of a set of users, which can confirm the reception using other of the buttons. The system not only provides locating features, but also scheduling and automation functionalities. The system can be easily integrated with a wide range of sensors and actuators using the variety of communication ports included in the n-Core Sirius A devices. By means of the automation engine provided by the n-Core API, the n-Core Polaris system can schedule automation tasks, as well as monitor all sensors in the environment in real-time.

4 Experiments and Results

The n-Core Polaris indoor locating system has been awarded as the winner of the first international competition on indoor localization and tracking, organized by the Ambient-Assisted Living Open Association (AALOA) [10] and performed in the Experimental Research Center in Applications and Services for Ambient Intelligence (CIAMI) sited in the Technological Park of Valencia (Spain) from 27th to 29th July 2011. Among the competitors there were companies and research groups coming from all Europe, including Germany, Austria, France, Switzerland, Ukraine and Spain. Finally, the results were presented in Lecce (Italy) within the framework of the AAL Forum from 26th to 28th September 2011.

In order to evaluate the competing localization systems, the following evaluation criteria were applied [10]. Each criterion had a maximum of 10 points. To calculate the overall score, each criterion was multiplied by a certain weight:

- *Accuracy (weight 0.25)*: each produced location sample was compared with the reference position, calculating the distance error. The final score on accuracy was the average between the scores obtained in the next two phases:
  - *Phase 1*: After a random walk the user stopped 30s in each Area of Interest (AoI). Accuracy was measured as the fraction of time in which the locating system provides the correct information.
  - *Phase 2*: The stream produced by competing systems was compared against a *logfile* of the expected position of the user. Specifically, the individual error of each measure was evaluated, and the 75th percentile of the errors was estimated. In this sense, In Figures 3, 4 and 5 can be seen the performance of n-Core Polaris in this phase, which achieves a 0.97m mean distance error.
- *Installation complexity (0.2)*: a measure of the effort required to install the AAL locating system in a $70m^2$ flat, measured by the evaluation committee as the total number of man-minutes of work needed to complete the installation. In this sense, the n-Core Polaris system was deployed in less than seven minutes in the flat, which demonstrates the ease of its installation.
- *User acceptance (0.2)*: expresses how much the locating system is invasive in the user's daily life and thereby the impact perceived by the user; this parameter is qualitative and was evaluated by the evaluation committee.
- *Availability (0.15)*: fraction of time the locating system was active and responsive. It was measured as the ratio between the number of produced location data and the number of expected data (one sample every half a second).
- *Integrability into AAL systems (0.1)*: use of open source solutions, use of standards, availability of developing libraries, integration with standard protocols.

As can be seen in Table 1, the n-Core Polaris system obtained the first place. These results demonstrate n-Core Polaris is a robust system suitable to be used in indoor environments, such as homes, hospitals or offices, and that can locate users and assets with up to 1m accuracy without interfering in the daily-life of people.

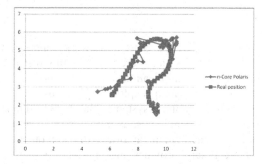

Fig. 3 Accuracy evaluation, phase 2, route 1 (Mean error = 0.777m; 3rd quartile = 1.056m).

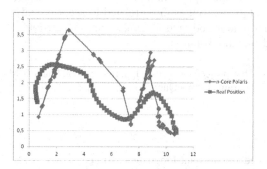

Fig. 4 Accuracy evaluation, phase 2, route 2 (Mean error = 1.055m; 3rd quartile = 1.306m).

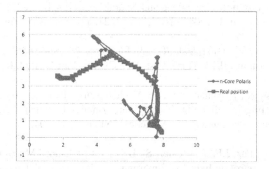

Fig. 5 Accuracy evaluation, phase 2, route 3 (Mean error = 1.088m; 3rd quartile = 1.338m).

Table 1 Intermediate and overall scores of the three best competitors in EvAAL Competition.

Competitor	Accuracy	Availability	Installation Complexity	User Acceptance	Integrability in AAL	Overall Score
n-Core Polaris	5.9611	9.8756	10	7.625	6.5	7.14
AIT team	8.4540	1.3674	6.82	6.875	8.5	5.90
iLoc	7.8007	9.3922	0	5.875	4.5	4.98

5 Conclusions

Among the wide range of Wireless Sensor Networks applications, Real-Time Locating Systems are emerging as one of the most exciting research areas. Health-care, surveillance or work safety applications are only some examples of the possible environments where RTLSs can be exploited. There also are different wireless technologies that can be used on these systems. The ZigBee standard offers interesting features over the rest technologies, as it allows the use of large mesh networks of low-power devices and the integration with many other applications.

In this regard, n-Core Polaris provides an important competitive advantage to applications where it is necessary to know the location of people, animals or objects. Amongst its multiple application areas are the healthcare, the industrial or the agricultural sectors, as well as those related to security and Ambient Assisted Living. Its optimal indoor and outdoor functioning makes n-Core Polaris a flexible, powerful and versatile solution.

Acknowledgments. This project has been supported by the Spanish Ministry of Science and Innovation (Subprograma Torres Quevedo).

References

1. Ambient Assisted Living Joint Programme, http://www.aal-europe.eu (accessed October 2011)
2. Nerguizian, C., Despins, C., Affès, S.: Indoor Geolocation with Received Signal Strength Fingerprinting Technique and Neural Networks. In: de Souza, J.N., Dini, P., Lorenz, P. (eds.) ICT 2004. LNCS, vol. 3124, pp. 866–875. Springer, Heidelberg (2004)
3. Liu, H., Darabi, H., Banerjee, P., Liu, J.: Survey of Wireless Indoor Positioning Techniques and Systems. IEEE Trans. Syst. Man Cybern. Part C-Appl. Rev. 37, 1067–1080 (2007)
4. Kaemarungsi, K., Krishnamurthy, P.: Modeling of indoor positioning systems based on location fingerprinting. In: Twenty-third Annual Joint Conference of the IEEE Computer and Communications Societies, INFOCOM 2004, vol. 2, pp. 1012–1022 (2004)
5. Tapia, D.I., de Paz, J.F., Rodríguez, S., Bajo, J., Corchado, J.M.: Multi-Agent System for Security Control on Industrial Environments. International Transactions on System Science and Applications Journal 4(3), 222–226 (2008)
6. Ding, B., Chen, L., Chen, D., Yuan, H.: Application of RTLS in Warehouse Management Based on RFID and Wi-Fi. In: 4th International Conference on Wireless Communications, Networking and Mobile Computing, WiCOM 2008, pp. 1–5 (2008)
7. Stelios, M.A., Nick, A.D., Effie, M.T., et al.: An indoor localization platform for ambient assisted living using UWB. In: Proceedings of the 6th International Conference on Advances in Mobile Computing and Multimedia, pp. 178–182. ACM, Linz (2008)
8. Nebusens: n-Core®: A Faster and Easier Way to Create Wireless Sensor Networks (2011), http://www.n-core.info (accessed October 2011)

9. Tapia, D.I., Alonso, R.S., Rodríguez, S., de la Prieta, F., Corchado, J.M., Bajo, J.: Implementing a Real-Time Locating System Based on Wireless Sensor Networks and Artificial Neural Networks to Mitigate the Multipath Effect. In: 2011 Proceedings of the 14th International Conference on Information Fusion (FUSION), pp. 1–8. IEEE/ISIF, Chicago, USA (2011)
10. AAL Open Association. Evaluating AAL Systems through Competitive Benchmarking (2011), http://evaal.aaloa.org (accessed October 2011)

Cyclic Scheduling for Supply Chain Network

Grzegorz Bocewicz, Robert Wójcik, and Zbigniew Banaszak

Abstract. This paper concerns the domain of Supply Chain Network Infrastructure (SCNI) usually observed in the multimodal transportation systems such as Multi-modal Passenger Transport Systems supported by lines of buses, trains, etc., and focuses on the scheduling problems encountered in these systems. SCNI can be modeled as a network of lines providing cyclic routes for particular kinds of stream-like moving transportation means. Lines and using them passengers can be seen as a multi agent system where passengers expectations compete with lines capability. The main question regards of SCNI schedulability, e.g. the guarantee the same distances in assumed different directions will require similar amount of the travel time. The declarative model of SCNI enabling to formulate cyclic scheduling problem in terms of the constraint satisfaction is our contribution.

Keywords: Cyclic scheduling, supply chain network, declarative modeling, multimodal process, constraints programming.

1 Introduction

A cyclic schedule [2], [8] is one in which the same sequence of states is repeated over and over again. In the case of Multimodal Transportation Systems (MTS) the

Grzegorz Bocewicz
Dept. of Computer Science and Management, Koszalin University of Technology,
Sniadeckich 2, 75-453 Koszalin, Poland
e-mail: bocewicz@ie.tu.koszalin.pl

Robert Wójcik
Wrocław University of Technology, Institute of Computer Engineering,
Control and Robotics, Wrocław, Poland
e-mail: robert.wojcik@pwr.wroc.pl

Zbigniew Banaszak
Dept. of Business Informatics, Warsaw University of Technology,
Narbutta 85, 02-524 Warsaw, Poland
e-mail: Z.Banaszak@wz.pw.edu.pl

J.M.C. Rodríguez et al. (Eds.): Trends in PAAMS, AISC 157, pp. 39–47.
springerlink.com   © Springer-Verlag Berlin Heidelberg 2012

appropriate cyclic scheduling problem has to take into account the constraints implied by the considered Supply Chain Network Infrastructure (SCNI), e.g. see Fig. 1. Assuming the transportation lines considered are cyclic and connected by common shared change stations a network can be modeled in terms of Cyclic Concurrent Process System (SCCP) [2]. Assuming each line is serviced by a set of stream-like moving transportation means (vehicles) and operation times required for traveling between subsequent stations as well as semaphores ensuring vehicles mutual exclusion on shared stations are given, the main question regards of SCNI timetabling, for instance guaranteeing the shortest time of the trip for passengers following a given direction. Depending on SCNI timetabling the time of the trip of passengers following different itineraries may dramatically differ. In that context the considered cyclic scheduling directly regards of multimodal processes encompassing passengers' itineraries, and indirectly regards of modeling them SCCPs. In systems of that type local lines (including transportation means) play the role of agents [1], attempting to reach their goals while following expectations of multi-modal processes. So, the considered MTS are treated as multi agent ones. The SCNI schedules sought have to follow vehicles collision- and deadlock-free flows as well as the passengers' itinerary optimization requirements. The problem considered belongs to NP-hard ones [3].

Literature Review. So far there is no research paper on cyclic scheduling of multimodal processes modeled in terms of above defined SCNI. The existing approach to solving the SCCPs scheduling problem base upon the simulation models, e.g. the Petri nets [5], the algebraic models, e.g. upon the (max,+) algebra [4] or the artificial intelligent methods [6]. The SCCP driven models assuming a unique process execution along each cyclic route, studied in [1], [2], [4] do not allow to take in to account the stream-like flow of local cyclic processes, e.g. buses servicing a given city line. So, this work can be seen as a continuation of the investigations conducted in [1], [2], [4], [7].

New Contributions. The declarative models employing the constraints programming techniques implemented in modern platforms such as OzMozart, ILOG, [1], [2] seems to be well suited to coup with SCNI scheduling problems. In that context, our contribution is a formulation of SCNI cyclic scheduling problem in terms of the constraint satisfaction one [2].

Organization. The paper is organized as follows. In Section 2, an illustrative example of SCNI and its cyclic scheduling problem statement are provided. In Section 3, a cyclic processes network is modeled. In Section 4, the selected case of multimodal processes is discussed In Section 5, we draw the conclusion.

2 Problem Formulation

The SCNI with distinguished vehicles and stations, shown in Fig. 1, is modeled in terms of the SCCP shown in Fig. 2. Four local **cyclic processes** (agents) are considered: P_1, P_2, P_3, P_4. The processes follow the **routes** (composed of transportation sectors and separating them stations) and while providing connections in two directions i.e., the north-south and the east-west, for **the two multimodal processes** (agents) mP_1, mP_2 and mP_3, mP_4, respectively. P_1, P_2 contain two sub-processes $P_1 = \{P_1^1, P_1^2\}, P_2 = \{P_2^1, P_2^2\}$ representing trains moving along the same route. The following constraints determine the processes cooperation:

- The new local process operation (the train's operation such as: passengers' transportation, boarding etc.) may begin only if the current operation has been completed and the resource designed to this operation is not occupied.
- The local processes share the common resources (the stations) in the mutual exclusion mode. The new local process operation can be suspended only if designed resource is occupied. The local processes suspended cannot be released. Local processes are non-preempted.
- The multimodal processes follow the local transportation routes. Different multimodal processes can be executed simultaneously along a local process.
- The local and multimodal processes are executed cyclically, resources occurring in each transportation route cannot repeat.

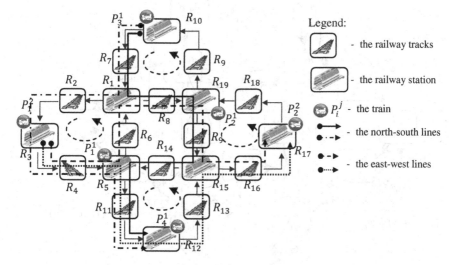

Fig. 1 An example of the SCNI

The main question concerns of SCCP cyclic steady state behavior and a way this state depends on direction of local process routes as well as on priority rules, and an initial process allocation to the system resources. Assuming the steady state there exists the next question regards of travel time along assumed multimodal process route linking distinguished destination points. Of course, the periodicity of multimodal processes depends on SCCP periodicity, i.e. characteristics of a given SCNI. That means an initial state and a set of dispatching rules can be seen as control variables allowing one to "adjust" multimodal processes schedule.

Consider a SCCP model of SCNI specified by the given dispatching rules, operation times (see Table 1), and initial processes allocation. The main question concerns: Does there exist a cyclic steady state of local and multimodal processes?

Table 1 Operation times of SCCP's o (from Fig. 2)

Streams	i	k	$t_{i,1}^k$	$t_{i,2}^k$	$t_{i,3}^k$	$t_{i,4}^k$	$t_{i,5}^k$	$t_{i,6}^k$
P_1^1	1	1	1	1	1	2	1	3
P_1^2	1	2	1	3	1	1	1	1
P_2^1	2	1	1	1	1	3	1	1
P_2^2	2	2	1	3	1	1	1	1
P_3^1	3	1	1	3	1	1	1	3
P_4^1	4	1	1	2	1	1	1	4

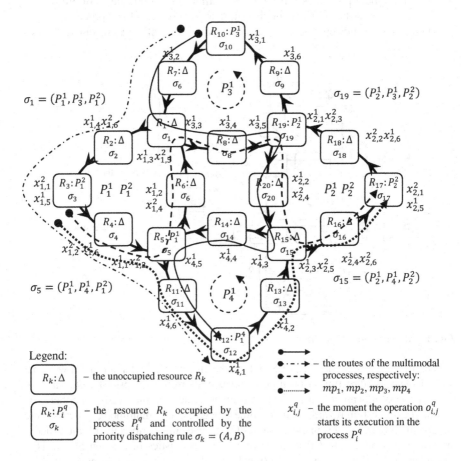

Fig. 2 SCCP of SCNI from Fig. 1

3 Modeling of Cyclic Processes Network

In the SCCP model of SCNI the following **notations** are used [1], [2]:

- A sequence $p_i^k = (p_{i,1}^k, p_{i,2}^k, \ldots, p_{i,lr(i)}^k)$ specifies **the route of the local process's stream P_i^k** (k-th stream of the i-th local process P_i), and its components define the resources used in course of process operations execution, where: $p_{i,j}^k \in R$ (the set of resources: $R = \{R_1, R_2, \ldots, R_m\}$) – denotes the resource used by the k-th stream of i-th local process in the j-th operation; in the rest of the paper **the j-th operation executed on resource $p_{i,j}^k$ in the stream P_i^k** will be denoted by $o_{i,j}^k$; $lr(i)$ - denotes a length of cyclic process route.

- $t_i^k = (t_{i,1}^k, t_{i,2}^k, \ldots, t_{i,lr(i)}^k)$ specifies **the process operation times**, where $t_{i,j}^k$ denotes the time of execution of operation $o_{i,j}^k$ (see Table 1).

- $mp_i = \left(mpr_j(a_j, b_j), mpr_l(a_l, b_l), \ldots, mpr_h(a_h, b_h) \right)$ specifies **the route of the multimodal process mP_i** where: $mpr_j(a, b) = \left(crd_a p_j^k, crd_{a+1} p_j^k, \ldots, crd_b p_j^k \right), crd_i D = d_i$, for $D = (d_1, d_2, \ldots, d_i, \ldots, d_w), \forall a \in \{1, 2, \ldots, lr(i)\}, \forall j \in \{1, 2, \ldots, n\}, crd_a p_j \in R$.
  The transportation route mp_i is a sequence of sections of local process routes. For the sake of simplicity let as assume the all operation times of multimodal processes are the same and equal to the 1 unit of time.

- $\Theta = \{\sigma_1, \sigma_2, \ldots, \sigma_m\}$ is the set of **the priority dispatching rules**, where $\sigma_i = (s_{i,1}, \ldots, s_{i,lp(i)})$ is the sequence components of which determine an order in which the processes can be executed on the resource R_i, $s_{i,j} \in P$ (the set of process streams: $P = \{P_1^1, \ldots, P_1^a, P_2^1 \ldots, P_2^b, \ldots, P_n^z\}$, each process executes periodically in infinity). Dispatching rules which determine an order on the shared train stations (resources R_1, R_5, R_{15}, R_{19}) are following: $\sigma_1 = (P_1^1, P_3^1, P_1^2), \sigma_5 = (P_1^1, P_4^1, P_1^2), \sigma_{15} = (P_2^1, P_4^1, P_2^2), \sigma_{19} = (P_2^1, P_3^1, P_2^2)$.

In that context a SCCP can be defined as a pair [2]:

$$SC = (SC_l, SC_m), \tag{1}$$

where: $SC_l = (R, P, \Pi, T, \Theta)$ – characterizes the SCCP structure, i.e.
$\qquad R = \{R_1, R_2, \ldots, R_m\}$ – the set of resources,
$\qquad P = \{P_1^1, \ldots, P_1^a, \ldots, P_n^1, \ldots, P_n^z\}$ – the set of local processes,
$\qquad \Pi = \{p_1, p_2, \ldots, p_n\}$ – the set of local process routes,
$\qquad T = \{T_1, \ldots, T_n\}$ – the set of local process operations times,
$\qquad \Theta = \{\sigma_1, \sigma_2, \ldots, \sigma_m\}$ – the set of dispatching priority rules.
$\qquad SC_m = (MP, M\Pi)$ – characterizes the SCCP behavior, i.e.
$\qquad MP = \{mP_1, mP_2, \ldots, mP_u\}$ – the set of multimodal processes,
$\qquad M\Pi = \{mp_1, mp_2, \ldots, mp_u\}$ – the set of multimodal process routes,

The main question concerns of SCCP cyclic behavior and a way this behavior depends on direction of local transportation routes Π, the priority rules Θ, and a set of initial states, i.e., an initial processes allocations to the system resources.

CSP-driven cyclic scheduling: Since parameters describing the SCCP model (1) are usually discrete, and linking them relations can be seen as constraints, hence related to them cyclic scheduling problems can be presented in the form of the Constraint Satisfaction Problem (CSP) [1], [2]. More formally, CSP is a framework for solving combinatorial problems specified by pairs: (a set of variables and associated domains, a set of constraints restricting the possible combinations of the variable values). The CSP relevant to the SCCP can be stated as follows [2]:

$$CS = ((\{R, P, \Pi, T, \Theta, X, Tc\}, \{D_R, D_\Pi, D_T, D_\Theta, D_X, D_{Tc}\}), C), \tag{2}$$

where:

- R, P, Π, T, Θ are the decision variables describing the structure of the SCCP, i.e., (1), and X, Tc are the decision variables describing the cyclic behavior of the SCCP. $X = \{X_1^1, \dots, X_1^a, X_2^1 \dots, X_2^b, \dots, X_n^z\}$ is the set of sequences $X_i^k = (x_{i,1}^k, x_{i,2}^k, \dots, x_{i,lr(i)}^k)$, where each variable $x_{i,j}^k$ determines **the moment of** $o_{i,j}^k$ **operation beginning** in any (the l-th) cycle: $x_{i,j}^k(l) = x_{i,j}^k + l \cdot Tc$, $l \in \mathbb{Z}$, (where $x_{i,j}^k(l) \in \mathbb{Z}$ – means the moment the $o_{i,j}^k$ operation starts its execution in the l-th cycle) and Tc is the SCCP periodicity: $Tc = x_{i,j}^k(l+1) - x_{i,j}^k(l)$.

- the domains $D_R, D_P, D_\Pi, D_T, D_\Theta, D_X, D_{Tc}$ of decision variables which describe the family of: the set of resources, set of processes, sets of admissible routings, sets of admissible operation times, sets of admissible dispatching priority rules, sets of admissible coordinate values X_i^k, $x_{i,j}^k \in \mathbb{Z}$, set of admissible values of variables Tc, respectively.

- the constraints determining the relationship between the structure (specified by the quin-tuple (R, P, Π, T, Θ)) and the behavior following from this structure (specified by (X, Tc)) can be defined by the operator max [2]. The constraints following assumptions imply for instance that an operation from the process P_1^1 can begin at the moment $x_{1,3}^1$ on resource R_1 only if the previous operation executed on the resource R_6 was completed at $x_{1,2}^1 + t_{1,2}^1$ and the resource R_1 has been released, i.e. if the process P_1^2 occupying the resource R_1 begins its subsequent operation at $x_{1,6}^2 - Tc + 1$. Therefore $x_{1,3}^1 = max(x_{1,6}^2 - Tc + 1; x_{1,2}^1 + t_{1,2}^1)$. The rest of operation starting moments can be determined by analogy, see Table 2

- **The system's cyclic behavior** encompasses itself through values of decision variables X, guaranteeing its periodicity Tc. The parameters determining the cyclic behavior such as X and Tc are solution to the problem (2) following the set of constraints C (Table 2.), determining the SCCP's structure (1).

4 Cyclic Processes Scheduling

Consider CSP stated by CS (2) and formulated for SCCP from Fig 2. The assumed set $\{\sigma_1 = (P_1^1, P_3^1, P_1^2),\ \sigma_{19} = (P_2^1, P_3^1, P_2^2),\ \sigma_5 = (P_1^1, P_4^1, P_1^2),\ \sigma_{15} = (P_2^1, P_4^1, P_2^2)\}$ of dispatching rules implies $Tc = 11$. The resultant cyclic steady state shown in Fig. 3 has be obtained in OzMozart, Dual Core 2.67, GHz, 2.0, GB RAM environment in 1 s. Obtained periodicity $(Tc = 11)$ of the SCNI behavior implies different traveling times required by different directions – the itineraries mP_4 and mP_3 following the routes mp_3, mp_4 along the east-west direction are realized in 18 and 28 time units, respectively (see the dotted and dashed lines in Fig. 1÷3). In turn, the itineraries mP_1 and mP_2 following the routes mp_1, mp_2 along the north-south direction are realized in 22 and 33 time units, respectively (see the solid and dot-dashed lines in Fig. 1÷3). So, the best line serving the east-west direction is faster than the best line serving the north-south direction.

Table 2 The constraints describing the moments $x_{i,j}^k$ of SCCP from Fig. 2

R_3:	$x_{1,1}^1 = \max\left(x_{1,6}^1 - Tc + 1; x_{1,6}^2 - Tc + t_{1,6}^2\right)$	R_4: $\quad x_{1,2}^2 = \max\left(x_{1,1}^1 + 1; x_{1,1}^2 + t_{1,1}^2\right)$
	$x_{1,5}^1 = \max\left(x_{1,2}^2 + 1;\ x_{1,4}^1 + t_{1,4}^1\right)$	$x_{1,6}^1 = \max\left(x_{1,3}^2 + 1; x_{1,5}^1 + t_{1,5}^1\right)$
R_5:	$x_{1,1}^1 = \max\left(x_{1,4}^2 - Tc + 1; x_{1,6}^1 - Tc + t_{1,6}^1\right)$	R_6: $\quad x_{1,2}^1 = \max\left(x_{1,5}^2 - Tc + 1; x_{1,1}^1\right.$
	$x_{4,5}^1 = \max\left(x_{1,2}^1 + 1; x_{4,4}^1 + t_{4,4}^1\right)$	$\left. + t_{1,1}^1\right)$
	$x_{1,3}^2 = \max\left(x_{4,6}^1 + 1;\ x_{1,2}^2 + t_{1,2}^2\right)$	$x_{1,4}^2 = \max\left(x_{1,3}^1 + 1; x_{1,3}^2 + t_{1,3}^2\right)$
R_1:	$x_{1,3}^1 = \max\left(x_{1,6}^2 - Tc + 1; x_{1,2}^1 + t_{1,2}^1\right)$	R_2: $\quad x_{1,4}^1 = \max\left(x_{1,1}^2 + 1;\ x_{1,3}^1 + t_{1,3}^1\right)$
	$x_{3,3}^1 = \max\left(x_{1,4}^1 + 1; x_{3,2}^1 + t_{3,2}^1\right)$	$x_{1,6}^2 = \max\left(x_{1,5}^1 + 1;\ x_{1,5}^2 + t_{1,5}^2\right)$
R_{17}:	$x_{2,1}^2 = \max\left(x_{2,6}^1 - Tc + 1; x_{2,6}^2 - Tc + t_{2,6}^2\right)$	R_{18}: $\quad x_{2,6}^1 = \max\left(x_{2,3}^2 + 1; x_{2,5}^1 + t_{2,5}^1\right)$
	$x_{2,5}^1 = \max\left(x_{2,2}^2 + 1; x_{2,4}^1 + t_{2,4}^1\right)$	$x_{2,2}^2 = \max\left(x_{2,1}^1 + 1; x_{2,1}^2 + t_{2,1}^2\right)$
R_{19}:	$x_{2,1}^1 = \max\left(x_{2,4}^2 - Tc + 1; x_{2,6}^1 - Tc + t_{2,6}^1\right)$	R_{20}: $\quad x_{2,2}^1 = \max\left(x_{2,5}^2 - Tc + 1; x_{2,1}^1\right.$
	$x_{3,5}^1 = \max\left(x_{2,2}^1 + 1; x_{3,4}^1 + t_{3,4}^1\right)$	$\left. + t_{2,1}^1\right)$
	$x_{2,3}^2 = \max\left(x_{2,6}^1 + 1;\ x_{2,2}^2 + t_{2,2}^2\right)$	$x_{2,4}^2 = \max\left(x_{2,3}^1 + 1; x_{2,3}^2 + t_{2,3}^2\right)$
R_{15}:	$x_{2,3}^1 = \max\left(x_{2,6}^2 - Tc + 1; x_{2,2}^1 + t_{2,2}^1\right)$	R_{16}: $\quad x_{2,4}^1 = \max\left(x_{2,1}^2 + 1; x_{2,3}^1 + t_{2,3}^1\right)$
	$x_{4,3}^1 = \max\left(x_{2,4}^1 + 1;\ x_{4,2}^1 + t_{4,2}^1\right)$	$x_{2,6}^2 = \max\left(x_{2,5}^1 + 1; x_{2,5}^2 + t_{2,5}^2\right)$
R_{12}:	$x_{4,1}^1 = x_{4,6}^1 - Tc + t_{4,6}^1$	R_{13}: $\quad x_{4,2}^1 = x_{4,1}^1 + t_{4,1}^1$
R_{14}:	$x_{4,4}^1 = x_{4,3}^1 + t_{4,3}^1$	R_{11}: $\quad x_{4,6}^1 = x_{4,5}^1 + t_{4,5}^1$
R_{10}:	$x_{3,1}^1 = x_{3,6}^1 - Tc + t_{3,6}^1$	R_7: $\quad x_{3,2}^1 = x_{3,1}^1 + t_{3,1}^1$
R_8:	$x_{3,4}^1 = x_{3,3}^1 + t_{3,3}^1$	R_9: $\quad x_{3,6}^1 = x_{3,5}^1 + t_{3,5}^1$

However, replacing the above assumed set of dispatching rules for the following new one $\{\sigma_1 = (P_3^1, P_1^1, P_1^2),\ \sigma_{19} = (P_2^1, P_3^1, P_2^2),\ \sigma_5 = (P_1^1, P_4^1, P_1^2),\ \sigma_{15} = (P_4^1, P_2^1, P_2^2)\}$ provides shorter cycle time $Tc = 10$, resulting in shortening of the travel time (20 time units) following the route mp_1 (north-south line), and extension of the travel time (28 time units) following the route mp_3 (east-west line).

That means, the different sets of dispatching rules implies different traveling times in assumed directions. In the case considered the difference between the shortest traveling times along two directions changes from $4 = 22 - 18$ to

$8 = 28 - 20$ time units. The open question is whether there exists such a set of dispatching rules guaranteeing the same best traveling time in both directions?

5 Concluding Remarks

In contradiction to the traditionally offered solutions the approach presented allows one to take into account such behavioral features as transient periods and deadlock occurrence. So, the novelty of the modeling framework lies in the declarative approach to reachability problems enabling an evaluation of multimodal cyclic process executed within cyclic processes environments. The approach presented leads to solutions allowing the designer to compose and synchronize elementary systems in such a way as to obtain the final SCNI system with required quantitative and qualitative behavioral features. So, we are looking for a method allowing one to replace the exhaustive search for the admissible control of the whole system by its a step-by-step structural design guaranteeing the required behavior (i.e., encompassing execution of assumed multi-modal processes).

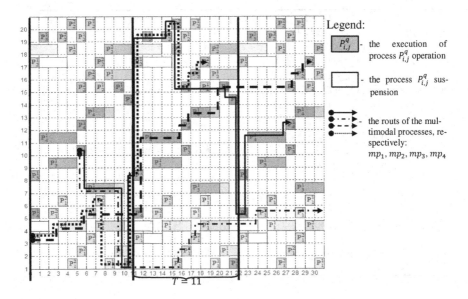

Fig. 3 Gantt diagram

References

[1] Bocewicz, G., Wójcik, R., Banaszak, Z.: Design of Admissible Schedules for AGV Systems with Constraints: A Logic-Algebraic Approach. In: Nguyen, N.T., Grzech, A., Howlett, R.J., Jain, L.C. (eds.) KES-AMSTA 2007. LNCS (LNAI), vol. 4496, pp. 578–587. Springer, Heidelberg (2007)

[2] Bocewicz, G., Banaszak, Z.: Declarative approach to cyclic scheduling of multimodal processes. In: Golińska, P. (ed.) EcoProduction and Logistics, vol. 1. Springer, Heidelberg (in print, 2012)

[3] Levner, E., Kats, V., Alcaide, D., Pablo, L., Cheng, T.C.E.: Complexity of cyclic scheduling problems: A state-of-the-art survey. Computers & Industrial Engineering 59(2), 352–361 (2010)

[4] Polak, M., Majdzik, P., Banaszak, Z., Wójcik, R.: The performance evaluation tool for automated prototyping of concurrent cyclic processes. Fundamenta Informaticae 60(1-4), 269–289 (2004)

[5] Song, J.-S., Lee, T.-E.: Petri net modeling and scheduling for cyclic job shops with blocking. Computers & Industrial Engineering 34(2), 281–295 (1998)

[6] Heo, S.-K., Lee, K.-H., Lee, H.-K., Lee, I.-B., Park, J.H.: A New Algorithm for Cyclic Scheduling and Design of Multipurpose Batch Plants. Ind. Eng. Chem. Res. 42(4), 836–846 (2003)

[7] Wang, B., Yang, H., Zhang, Z.-H.: Research on the train operation plan of the Beijing-Tianjin inter-city railway based on periodic train diagrams. Tiedao Xuebao/Journal of the China Railway Society 29(2), 8–13 (2007)

[8] Von Kampmeyer, T.: Cyclic scheduling problems, Ph.D. Dissertation, Fachbereich Mathematik/Informatik, Universität Osnabrück (2006)

Improving Production in Small and Medium Enterprises

María L. Borrajo, Javier Bajo, and Juan F. De Paz

Abstract. Knowledge management has gained relevance during the last years to improve business functioning. However, there is still a growing need of developing innovative tools that can help small to medium sized enterprises to detect and predict undesired situations. This article present a multi-agent system aimed at detecting risky situations. The multi-agent system incorporates models for reasoning and makes predictions using case-based reasoning. The models are used to detect risky situations and an providing decision support facilities. An initial prototype was developed and the results obtained related to small and medium enterprises in a real scenario are presented.

Keywords: Hybrid neural intelligent system, CBR, MAS, Business Intelligence, business risk prediction.

1 Introduction

Knowledge Management is a fundamental asset for businesses in the contemporary economy. Knowledge takes into account the organization of the

María L. Borrajo
Dept. Informática, University of Vigo, Campus Universitario As Lagoas s/n, Ourense, 32004, Spain
e-mail: lborrajo@uvigo.es

Javier Bajo
Facultad de Informática, Universidad Pontificia de Salamanca, Compañía 5, 37002, Salamanca, Spain
e-mail: jbajope@upsa.es

Juan F. De Paz
Departamento Informática y Automática
Universidad de Salamanca, Plaza de la Merced s/n, 37008, Salamanca, Spain
e-mail: fcofds@usal.es

J.M.C. Rodríguez et al. (Eds.): Trends in PAAMS, AISC 157, pp. 49–56.
springerlink.com   © Springer-Verlag Berlin Heidelberg 2012

business, individuals and the information [12]. Knowledge management can be applied to different organizations and different contexts. In the present financial context, it is increasilly relevant to provide innovative tools and decision support systems that can help the small-medium enterprises (SMEs) to improve their functioning [8], [11]. These tools and methods can contribute to improve the existing business control mechanisms, reducing the risk by predicting undesiderable situations and providing recommendations based on previous experiences [2].

This article presents an innovative approach, based on multi-agent systems [10], to propose a model for risk management and prediction in SMEs. Multi-agent systems are the most prevalent solution to construct Artificial Intelligence distributed systems. Intelligent agents can incorporate advanced artificial intelligence models to predict risky situations. In this study we propose a distributed approach where the components of a SME are modeled as intelligent agents that collaborate to create models that can evolve over the time and adapt to the changing conditions of the environment. Thus, making possible to detect risky situations for the SMEs and providing suggestions and recommendations that can help to avoid possible undesiderable situations. The core of the multi-agent system are the evaluator and advisor agents, that incorporate new techniques to analyze the data from enterprises, extract the relevant information, and detect possible failures or inefficiencies in the operation processes.

The article is structured as follows: the next section briefly introduces the problem that motivates this research. Section 3 presents the multi-agent system for managing small and medium enterprises and Section 4 describes its implementation. Section 5 presents the results obtained after testing the system.

2   Enterprise Risk Management

"Risk Management" is a broad term for the bussiness discipline that protects the assets and profits of an organization by reducing the potential for risks before it occurs, mitigating the impact of a loss if it occurs, and executing a swift recovery after a loss occurs. It involves a series of steps that include risk identification, the measurement and evaluation of exposures, exposure reduction or elimination, risk reporting, and risk transfer and/or financing for losses that may occur. All organizations practice risk management in multiple forms, depending on the exposure being addressed [1].

The economic environment has increased the pressure on all companies to address risks at the highest levels of the organization. Companies that incorporate a strategic approach to risk management use specialized tools and have more structured and frequent reporting on risk management. As such, they are in a better position to ensure that risk management provides relevant and applicable information that meets the needs of the organization and executive team. But no matter what an organization's approach is, the tools used must be backed up by solid, actionable reporting addressed [1]. It's not always necessary for the risk managers to be conducting their own studies for their voices to be heard. Forging a strong relationship with internal auditors and other departments can allow risk practitioners to supplement their reports with the risk manager's own analysis [3]. Enterprise Risk Management (ERM) is defined as "a process, effected by an entity's board of directors, management and other personnel, applied in strategy-setting and across the enterprise, designed to identify potential events that may affect the entity, and manage risk to be within its risk appetite, to provide reasonable assurance regarding the achievement of entity objectives." [4]. The managing of risks and uncertainties is central to the survival and performance of organizations. Enterprise risk management (ERM) is an emerging approach to managing risks across different business functions in an organisation that represents a paradigm shift from specialized approaches in managing specific risks [6], [7]. This paper provides a web intelligent model to ERM, which will subsequently lead to better organisational performance. ERM represents a revolutionary change in an organization's approach to risk. In addition, ERM encompasses all aspects of an organization in managing risks and seizing opportunities related to the achievement of the organization's objectives, not only for protection against losses, but for reducing uncertainties, thus enabling better performance against the organization's objectives [1].

3 Multi-agent System for Risk Management

In this article we propose a multi-agent system aimed at providing advanced capacities for risk management in SMEs. The multi-agent system provides a web system interface to facilitate the remote interaction with the human users involved in the risk management process. The core of the multi-agent system is a type of agent so called CBR-BDI agent. This agent type integrates a case-based reasoning mechanism (CBR) in its internal structure to take advantage of the reasoning abilities of the CBR paradigm. CBR-BDI agents are characterized by their capacities for learning and adaptation in dynamic environments. These agent types are used to evaluate the business' status and to generate recommendations that can help the business to avoid risky situations. CBR-BDI agents collaborate with other deliberative agents in the system to find optimum models for risk management. The agents in the system allow the users to access the system through distributed applications, which run on different types of devices and interfaces (e.g. computers, cell phones, PDA). Figure 1 shows the basic schema of the proposed architecture, where all requests and responses are handled by the agents in the platform. The system is modelled as a modular multi-agent architecture, where deliberative BDI agents are able to cooperate, propose solutions on very dynamic environments, and face real problems, even when they have a limited description of the problem and few resources available. These agents depend on beliefs, desires, intentions (BDI) and plan representations to solve problems. There are different kinds of agents in the architecture, each one with specific roles, capabilities and characteristics:

Business Agent. This agent was assigned for each firm in order to collect new data and allow consultations. The enterprise can interact with the system by means of this agent, introducing information and receiving predictions.

Evaluator Agent. It is responsible for the evaluation and predictions of potential risky situations. Every time that it is necessary to obtain a new estimate of the state of an activity, the agent evolves through several phases. On the one hand, this evolution allows the multi-agent system, to identify the latest situations most similar to the current situation in the retrieval stage, and to adapt the current knowledge in the reuse stage in order to generate an initial estimate of the state of the activity being analysed. On the other hand, it is possible to identify old situations that serve as a basis to detect the inefficient processes developed within the activity and to select the best of all possible activities. The activity selected will then serve as a guide for establishing a risk level for the activity, its function, and the company itself, to develop in a more positive way. The retain phase guarantees that the system evolves in parallel with the firm, basing the corrective actions on the calculation of the error previously made.

Advisor agent. The objective of this agent is to carry out recommendations to help the internal auditor decide which actions to take in order to improve the company's internal and external processes.

Expert Agent. This agent helps the auditors and enterprise control experts that collaborate in the project to provide information and feedback to the multi-agent system. These experts generate prototypical cases from their experience and they receive assistance in developing the Store agent case-base.

Store Agent. This agent has a memory that has been fed with cases constructed with information provided by the enterprise (through its agent) and with prototypical cases identified by 34 enterprises control experts, using personal agents who have collaborated and supervised the developed model.

4   A Practical Implementation

The application of agents and multi-agent systems provides the opportunity of taking advantage of the inherent capabilities of the agents. Nevertheless, it is possible to increase the reasoning and learning capabilities by incorporating a CBR [9] mechanism into the agents. In the case at hand, we will focus on the CBR-BDI agents [10], responsible for classifying the enterprise situation and predict possible risks as well as providing recommendations to manage risk situations. In the BDI model, the internal structure of an agent and its capability to choose is based on mental aptitudes: agent behaviour is composed of beliefs, desires, and intentions [10]. Case-based Reasoning is a type of reasoning based on the use of past experiences [9]. The fundamental concept when working with case-based reasoning is the concept of case. A case can be defined as a past experience, and is composed of three elements: A problem description which describes the initial problem, a solution which provides the sequence of actions carried out in order to solve the problem, and the final state which describes the state achieved once the solution was applied. The way in which cases are managed is known as the case-based reasoning cycle. This CBR cycle consists of four sequential steps: retrieve, reuse, revise and retain. The retrieve phase starts when a new problem description is received. Similarity algorithms are applied in order to retrieve from the case's memory the cases with a problem description more similar to the current one. Once the most similar cases have been retrieved, in the reuse phase the solutions of the cases retrieved are adapted to obtain the best solution for the current case. The revise phase consists of an expert revision of the solution proposed. Finally, the retain phase allows the system to learn from the experiences obtained in the previous phases and updates the case memory in consequence.

The Evaluator and Advisor agent use the same type of case and share the same memory of cases. The data for the cases were obtained by surveys conducted with enterprise experts in the different functional areas of various enterprises, using the Expert agents. This type of survey attempts to reflect the experience of the experts in their different fields. For each activity, the survey presents two possible situations: the first one tries to reflect the situation of an activity with an incorrect activity state, and the second one tries to reflect the situation of an activity with a satisfactory activity state. Both situations will be evaluated by a human expert using a percentage. Each activity is composed of tasks, and each task has an importance rate, and values of realization for both incorrect and satisfactory activity state. These parameters are explained below in the analysis of the case structure. Each case is composed of the following attributes:

- Case number: Unique identification: positive integer number.
- Input vector: Information about the tasks (n sub-vectors) that constitute an industrial activity: $((IR1,V1),(IR2,V2),...,(IRn,Vn))$ for n tasks. Each task sub-vector has the following structure (IRi,Vi):
  - IRi: importance rate for this task within the activity. It can only take one of the following values: VHI (Very high importance) with a numeric value of 5, HI (High Importance) with a numeric value of 4, AI (Average Importance) with a numeric value of 3, LI (Low Importance) with a numeric value of 2, VLI (Very low importance) with a numeric value of 1.
  - Vi: Value of the realization state of a given task: a positive integer number (between 1 and 10).
- Function number: Unique identification number for each function
- Activity number: Unique identification number for each activity
- Reliability: Percentage of probability of success. It represents the percentage of success obtained using the case as a reference to generate recommendations.
- Activity State: degree of perfection for the development of the activity, expressed by percentage. This is the solution of a problem case.

The following sub-sections present the internal structure of the CBR-BDI Evaluator and Advisor agents used to predict and prevent crisis in SMEs.

5   Results

A case study aimed at providing innovative web business intelligence tools for the management of SMEs was carried out in the Castilla y León region, in Spain. The experiment consisted on the construction of the initial prototype of cases memory, predicting potential risky situations for the enterprises taken into consideration and providing recommendations. The case study presented in this work was oriented to detect possible risky situations in SMEs, taken into account the crisis that affects the market. A multi-agent system was implemented and 22 SMEs participated in the experiment and were assigned a personal business agent. The enterprises were situated in different sectors of the Spanish market. The system was tested during 24 months, from January 2008 to January 2010, tuned and improved taking into account the experience acquired using a total of 238 cases.

To validate the overall functioning of the system it was necessary to individually evaluate the Evaluator and Advisor agents. These agents provide predictions on the performance of the activities and detect those tasks that can be improved for each activity in order to get an overall improvement of the activity. In the following paragraphs we will focus on the evaluation of the CBR-BDI agents and their influence in the multi-agent system. To validate the performance of the Evaluator agent, an estimation of the efficiency of the predictions provided by this agent was carried out. To evaluate the significance of the different techniques integrated within the Evaluator agent, a cross validation was established, following the Dietterich's 5x2- Cross-Validation Paired t-Test algorithm [5]. The value 5 in the algorithm represents the number of replications of the training process and value 2 is the number of sets in which the global set is divided. Thus, for each of the techniques, the global dataset S was divided into two groups S1 and S2 as follows: $S = S_1 \cup S_2$ y $S_1 \cap S_2 = \varnothing$. Then, the learning and estimation processes were carried out. This process was repeated 5 times and had the following steps: the system was trained using S1 and then it was used to classify S1 y S2. In a second step, the system was trained using S2 and then it was used to classify S1 y S2. The results obtained by the evaluator agent using the mixture of experts, presented in section 4, were compared to the results obtained using an individual RBF and an individual MLP to the same dataset and the same 5x2 Cross-Validation process. Table 1 shows the error rate obtained for each of the techniques, using the test in each of the 5 repetitions. As can be seen in Table 1, the estimated error was lower for the Evaluator agent than for the rest of the evaluated techniques.

Table 1 Absolute error for the estimation of the status of the activities.

Method	S_2	S_1	S_2	S_1	S_2	S_1	S_2	S_1	S_2	S_1
Advisor agent	0.297	0.309	0.210	0.281	0.207	0.355	0.226	0.343	0.239	0.302
MLP	0.677	0.669	0.489	0.507	0.513	0.806	0.530	0.696	0.506	0.485
RBF	1.009	0.833	0.656	0.985	0.878	0.959	0.620	0.831	0.643	0.783

A Paired t-Test was applied to check that the difference between the methods can be considered as significant if a value $\alpha=0.05$ is established. To evaluate the Advisor agent it is necessary to take into account that the aim of this agent is to detect inefficient tasks by means of gain functions. The evaluation of the functioning of the Advisor agent was carried out by selecting those tasks with higher values for the gain function. The selected tasks were used to estimate the

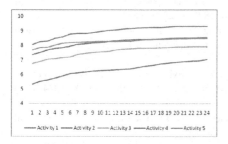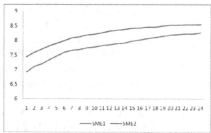

Fig. 1 a) Evolution of the average status of 5 activities during 12 months. b) Evolution of the average status of 2 SMEs during 12 months.

different scenarios for different execution values for the task. In this way, Figure 1a presents the evolution of the system for the average status of 5 activities along 12 months. As shown, the evolution for the 5 activities can be considered as positive. Looking at the evolution of the global efficiency for the activities analysed for two SMEs, shown in Figure 1b, it is possible to observe a growing tendency in the average status of the business along the time, which indicates a reduction of inefficient tasks in each of the activities. The results obtained demonstrate that the multi-agent system caused a positive evolution in all enterprises. This evolution was reflected in the reduction of inefficient processes. The indicator used to determine the positive evolution of the companies was the state of each of the activities analysed. After analysing one of the company's activities, it was necessary to prove that the state of the activity (valued between 1 and 100) had increased beyond the state obtained in the previous three month period. The system considers small changes in the tasks performed in the SMEs, and all the experts that participated in the experiments considered 3 months as a significant time to evaluate the evolution of a SME related to these changes.

Acknowledgments. This work has been supported by the Spanish Ministry of Science and Innovation project TIN2009-13839-C03.

References

[1]   Calderon, T.G., Cheh, J.J.: A roadmap for future neural networks research in auditing and risk assessment. International Journal of Accounting Information Systems 3(4), 203–236 (2002)

[2]   Chi-Jie, L., Tian-Shyug, L., Chih-Chou, C.: Financial time series forecasting using independent component analysis and support vector regression. Decision Support Systems 47(2), 115–125 (2009)

[3]   Colbert, J.L.: Risk. Internal Auditor, 36–40 (1995)

[4]   Committee of Sponsoring Organizations of the Treadway Commission (COSO), Guidance on Monitoring Internal Control Systems. COSO's Monitoring Guidance (2009)

[5] Dietterich, T.G.: Approximate statistical tests for comparing supervised classification learning algorithms. Neural Computation, 1895–1923 (1998)

[6] Ding, Y., Song, X., Zen, Y.: Forecasting financial condition of Chinese listed companies based on support vector machine. Expert Systems with Applications 34(4), 3081–3089 (2008)

[7] Huang, S., Tsai, C., Yen, D.C., Cheng, Y.: A hybrid financial analysis model for business failure prediction. Expert Systems with Applications 35(3), 1034–1040 (2008)

[8] Khashman, A.: A neural network model for credit risk evaluation. International Journal of Neural Systems 19(4), 285–294 (2009)

[9] Kolodner, J.: Case-Based Reasoning. Morgan Kaufmann, San Francisco (1983)

[10] Laza, R., Pavón, R., Corchado, J.M.: A Reasoning Model for CBR_BDI Agents Using an Adaptable Fuzzy Inference System. In: Conejo, R., Urretavizcaya, M., Pérez-de-la-Cruz, J.-L. (eds.) CAEPIA/TTIA 2003. LNCS (LNAI), vol. 3040, pp. 96–106. Springer, Heidelberg (2004)

[11] Sun, J., Li, H.: Financial distress prediction based on serial combination of multiple classifiers. Expert Systems with Applications 36(4), 8659–8666 (2009)

[12] Takeishi, A.: Knowledge Partitioning in the Interfirm Division of Labor: The Case of Automotive Product Development. Organization Science 13(3), 321–338 (2002)

Multiagent System for Detecting and Solving Design-Time Conflicts in Civil Infrastructure

Jaume Domínguez Faus, Francisco Grimaldo, and Fernando Barber

Abstract. One typical source of problems in the Civil Infrastructure domain is the distributed and collaborative nature of the projects in which different profiles of engineers contribute with designs devoted to the interest of their field of expertise. Thus, situations in which there are different conflicts of interests are quite common. A conflict refers to a situation in which the actions of an engineer collide with the interests of other engineers. In this paper, we present a multi-agent system that, thanks to the use of ontologies and rules on those ontologies, is able to detect profile-specific conflict situations and solve them according to the preferences of the parties involved in the conflict. The conflict solving is based on the Multi Agent Resource Allocation (MARA) theory. The system is applied to a real use case of an urban development where both the road network and the buildings are designed.

1 Introduction and Related Work

Interoperability is an often addressed term when enumerating the problems of distributed systems. In the same way that communication becomes difficult between two people speaking different languages, communication is difficult when dealing with systems relying on data for modeling a problem to solve. This happens because data only describe things and, as any description, they can be interpreted in many ways. Any infrastructure project as, for instance, a road construction involves lots of disciplines ranging from land-use to security regulations, with noise emission, road tracing, water drainage and many others in between.

Jaume Domínguez Faus
Centre for 3D GeoInformation, Aalborg University, 9220 Aalborg, Denmark
e-mail: jaume@land.aau.dk

Francisco Grimaldo · Fernando Barber
Departament d'Informàtica, Universitat de València, Av. de la Universitat s/n, (Burjassot) València, Spain 46100
e-mail: {francisco.grimaldo,fernando.barber}@uv.es

J.M.C. Rodríguez et al. (Eds.): Trends in PAAMS, AISC 157, pp. 57–64.
springerlink.com   © Springer-Verlag Berlin Heidelberg 2012

The general tendency is to have -when possible- a data model for each discipline that is used by specific software packages to assist the daily engineer's life and split the project works in discipline-experts teams. Problems arise when all the works done by different teams have to be put together. Issues like design clashes, synchronization problems, exceeded budgets, conflicts of interests appear, etc. as a consequence of the decentralized way of working with heterogeneous data models. The detection and solving of such problems is still a prominent manual work and some of them might remain undetected when this process is finished. Once the construction starts, the consequences of mistakes or suboptimal design cause that the infrastructure cost increases a 5-10% of the total budget in average [4]. Most of the efforts done so far have focused on avoiding the collisions by improving interoperability among different data models. It has not been, however, until recently when the conflict-solving has gained attention. This paper presents a new multiagent-based approach for detecting and solving design-time conflicts in the Civil Infrastructure domain. Currently, the Civil Infrastructure software industry focuses on making models that integrate more and more aspects of design disciplines in order to increase interoperability. Perhaps, the most advanced results of these efforts are the most successful standard files (such as CityGML or IFC [6], and Auto-CAD's DWG) or the Building Information Model (BIM) servers [4] which combine CAD models with management spreadsheets and other other documents to provide an integral project life-cycle management. However, this distributed and collaborative work has to deal with conflicts that inevitably appear when sub-designs of a project are merged.

Multi-agent Systems (MAS) have been suggested to aid in Civil Infrastructure projects. It is possible to find examples of MAS focused on controlling machinery [9], or on the distinct phases of a project: the tendering procedure [10]; the material supply chain [11]; and the construction phase [8] and [13]. Nevertheless, to the authors' knowledge there is a gap that has not yet been considered satisfactory: the negotiation between designer expertises in the Design phase of the project. Even though it is possible to find some problem-specific works like [1], the situated nature of this collaborative work makes the problem of abstraction of a system to be wicked [5]. However, this abstraction is necessary to capture the negotiation as a design conflict-solver in the software packages normally used by the engineers in their daily work. Thus, more research is needed in this field for MAS to be a real option.

The use of ontologies has been proposed as a means to give sense and semantics to the data in several contexts. In geospatial and civil infrastructure information, ontologies are not widely used. Although it is possible to envisage ontological structures in some data models (e.g. CityGML) they are hardly used in a formal and explicit manner. We propose the use of ontologies to support automatic conflict detection and of the Multi Agent Resource Allocation (MARA) [3] for its solving at a semantic level. In section 2.1 we present the ontological approach we propose to represent the world semantics, and the rules that are used to detect conflicts. Further below, in section 2.2 the negotiation mechanism used to solve conflicts is introduced. Finally, in section 3 we describe a use case in which the system was applied in order to illustrate its usefulness.

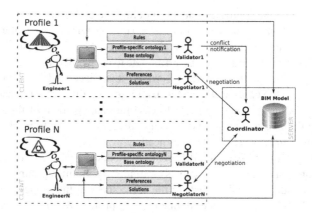

Fig. 1 Overview of the system

2 Architecture of the System

The multi-agent system we propose for detecting and solving design-time conflicts in the Civil Infrastructure domain is depicted in figure 1. It follows a distributed architecture approach allowing the engineers of different profiles to design, through their client interfaces, a common BIM model that is stored at the server. This collaborative work is carried out with the assistance of a set of agents: the Validators, the Negotiators and the Coordinator. The Validator agents are in charge of semantically detecting conflicts and errors within the model by using the ontological knowledge of each field of expertise. In turn, Negotiator agents aim at solving conflicts by expressing the preferences of the engineers in a negotiation protocol that is initiated by the Coordinator under conflict notification. Following, we review the details about these agents, which have been implemented as part of an agent society in JADE [2].

2.1 Semantic Conflict Detection

We propose using OWL [12] ontologies for the semantic abstraction of the data beyond the pure classical attribute/value pair. As shown in figure 1, our ontologies are structured in layers in which each layer provides an extra level of abstraction. At the lowest level, the Base Ontology defines the basic concepts needed by any geospatial data model. The base class `Feature` refers to the most basic object that traditionally forms geospatial data models such as GIS or CAD systems. A `Feature` is composed of a `Geometry` and of a set of `Attributes` defined by its name, its type, and its value. `Features` can be related to each other through the generic relation `hasRelationship` and its inverse relation `isRelationshipOf`. This pattern has proven to be flexible and suitable for many uses. Besides, by using inheritance, classes can be arranged in a hierarchy (e.g. `Conflict` and `Error` are particular types of `Problems`). Therefore, this ontology acts as the first layer of abstraction allowing the creation of Profile-specific ontologies on top of it.

Profile-specific Ontologies are meant to define the concepts of interest for each profile. These concepts can be specific `Features` providing particular properties and/or semantic meaning (e.g. a `Building` or a `Parcel`) and also specific relationships defining how certain `Features` relate to each other (e.g. `isLocatedAt` relates a `Building` with the `Parcel` where it is placed). This second level of ontologies allows to separate the categorization of the different interests involved in civil infrastructure projects in order to ease the management of the knowledge. Note, however, that this does not necessarily prevent a concept to be shared among different profiles in case several profiles need it.

As the project progresses, the different engineers include new designs or edit the existing ones and the model changes continuously. In this dynamic context, the Validator agent automatically assists in the correctness of the model as a whole by periodically checking a set of rules defined for the profile. We propose using SWRL [7] rules as a way of supporting the semantic consistency and ontological reasoning. These rules are ontological expressions with an antecedent and a consequent that allows the Validator agent to detect and infer problematic situations. The problems found are categorized between `Errors` or `Conflicts`, acording to the classes defined in the Base Ontology. `Errors` are situations in which the model is not correct due to missing or wrong values in the `Feature`'s properties and, thus, they are notified and solved manually by the engineer through its client interface. On the other hand, `Conflicts` capture the situations where the designers' interests collide and they are solved through the negotiation protocol explained next.

2.2   Conflict Solving Protocol

We propose using a Multi-Agent Resource Allocation[3] (MARA) approach to analyze the possible alternatives that solve the `Conflicts`. The MARA model provides agents with a general mechanism to make socially acceptable decisions. In this kind of decisions, members are required to express their preferences with regard to the different solutions that have been previously proposed by all the members for a specific decision problem. Our MARA approach uses ContractNet-like protocol as the allocation procedure. Figure 2 depicts the process for the case of a `Conflict` between two profiles. When the `Conflict` is detected by one Validator agent, it is notified to the Coordinator agent. Then the Coordinator distributes the `Conflict` to all the Negotiators in a Call For Proposals. The negotiators respond with their alternatives, if any, and the Coordinator collects all the proposals. In the collection, invalid or repeated solutions are filtered out and the set of remaining solutions is distributed again to request the preferences. Each Negotiator then expresses its utility on each of the solutions at hand by giving it a value ranging from 0 (lowest) to 10 (highest). The Coordinator agent then picks the winner solution which is the one that maximizes the global utilitarian social welfare represented by the solution that accumulates highest utility among the negotiators. Finally, the winner solution is then broadcasted to all the clients.

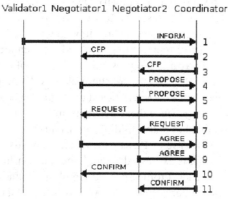

Fig. 2 JADE[2] console showing the Conflict Solving protocol

3 Urban Development Use Case

In order to simulate the daily work of engineers in the design phase, an urban development use case was selected. This project consists of the development of the Strømsø area in the city of Drammen, Norway. Traditionally an industrial area, after decades of growth, Strømsø became the downtown of Drammen while keeping the original industrial aspect. The authorities want to adapt it to the new residential reality. In general, the development goal is the construction of residential buildings to increase the number of inhabitants. In the initial phase, the project defines where to place buildings according to their characteristics (number of residents, floors and footprint). Further phases of the design deal with other more detailed aspects. We focused on the building placing problem to show our MAS approach.

To avoid future traffic jams, it was agreed that there should not be more residents than the capacity of the road. Thus, the location of a building is constrained to the capacity of the road that serves the building. The current usage of the road is obtained by the sum of the inhabitants of the buildings that are associated to that road. So, in addition to their geometry, buildings and roads specify the amount of inhabitants and the road capacity respectively in their attributes. There are two engineering profiles identified: 1) The designer that places buildings in a location of her/his choice (Building profile), and 2) The road designer that detects which road is connecting the building to the road network and checks whether the road is capable to hold all the buildings connected to it (Road profile).

For each profile there is a designer that is developing the model, and each designer has: 1) a validator agent that checks the model according to the semantics (expressed by her/his ontology and rule-set settings) and initiates the negotiation; and 2) a negotiator agent that performs the negotiation on behalf of the engineer.

As introduced in section 2.1, Features constitute the most generic object that can be defined in our ontological model (unlike pure geometry-based models in which basically only the geometry is known). On top of it, we identify three specific concepts of interest for this use case: the `Road`, the `Parcel` and the `Building`. `Parcels` are `Features` that define an area in which `Buildings` can be placed. `Buildings` are `Features` representing the residential entities where people live in. In turn, `Roads` are `Features` representing the parts of the road network. Beyond specifying the type of a `Feature`, these classes define more attributes that are required to describe their characteristics such as the capacity of a `Road` or the inhabitants of a `Building`. Since these concepts are relevant for both profiles, the previous classes are defined in both Building and Road Profile-specific Ontologies.

Since what a particular type of `Feature` means depends on the profile that is looking at it and, in turn, it is expressed through the relationships that it establishes with other `Features`, each Profile-specific Ontology defines a particular set of relationships the profile is interested in. The layered design of the ontologies allowed the Road profile to define the relationship called `roadServesTo` and its inverse `isServedByRoad` which state what `Road` serves any other `Feature` and vice versa. On the other hand, the Building profile defines the relationship `holds` and its inverse `isLocatedAt` which establishes which `Feature(s)` a given `Parcel` holds and, inversely, where a particular `Feature` is located.

SWRL rules have been defined for each profile so that the corresponding Validator agent can detect the errors and conflicts that appear in the model and that are related to its field of expertise. Regarding the errors, for example, each `Road` must specify its capacity in order to check if it can hold the potential traffic. If a `Road` is missing this attribute, then the model is not complete and the validator agent infers an `Error`. Equation 1 shows the rule used to infer a `RoadCapacityError`, a specific type of `Error` defined in the Profile-specific Ontology for the Road profile. This rule could be read as: if an element r happens to be a `Road`, and the result of the operation *isMissingAttribute* for this road and the attribute name "capacity" resolves to true, then r is also a `RoadCapacityError`. The operation *isMissingAttribute* is an example of how it is possible to extend the general logic operations of ontologies with user-defined operations. This mechanism is allowed in SWRL rules by means of the use of Built-ins. Similar rules were used by the Building profile to detect when a building does not declare the amount of inhabitants.

On the other hand, the two profiles involved in this use case may also come into conflict. That is the case when the building designer places a building in a parcel where the road connecting to that parcel cannot hold the new population of the building. Equation 2 shows the rule defined in the Building profile to detect this kind of `RoadExhaustedConflict`. This rule is actually a compound rule that: retrieves the `Parcel` p where the `Building` b is placed, gets the `Road` r serving that parcel and computes whether the road is overloaded with the buildings that are connected to it through the operation *isRoadExhausted*. This rule leans on the inference done by another rule about which road serves a parcel (equation 3). This latter rule explores the relationships of the ontology to get the `Geometry` gp of the `Parcel` p and selects the closest `Road` r by means of the operation *closestRoad*.

$$Road(?r) \land isMissingAttribute(?r, "capacity") \rightarrow RoadCapacityError(?r) \quad (1)$$

$$isLocatedAt(?b, ?p) \land isServedByRoad(?p, ?r) \land \rightarrow$$
$$RoadExhaustedConflict(?r, ?b)$$
$$Road(?r) \land isRoadExhausted(?r) \quad (2)$$

$$Parcel(?p) \land hasGeometry(?p, ?pg) \land \rightarrow isServedByRoad(?p, ?r)$$
$$closestRoad(?pg, ?r) \quad (3)$$

The Conflict going to be solved is detected by the Validator agent (see figure 1) who provokes the initiation of the conflict solving protocol described above and depicted in 2. When the Coordinator is notified, he broadcasts the Call For Proposals to the Negotiators acting as proxies of the engineers. The engineers at the clients receive a message informing that a new Conflict solving sequence has been started and the Coordinator is waiting for their proposals. Knowing the details of the conflict (i.e. the Road is exhausted) they think on how to fix it. Different solutions like, e.g., enlarging the Road for more capacity; or reducing the amount of inhabitants of one or several Buildings; or maybe relocating a Building in another Parcel served by a Road with more availability for new residents; are then applied temporarily to the model. A recording system allows to capture the changes to the model. The engineers encapsulate sequences of changes (such as "on Building number 32, set the value of the Attribute 'inhabitants' to 30 from 40") into a Solution and provide all the alternative Solutions they have. All the alternatives are then proposed to the Coordinator who evaluates them and discards repeated or invalid ones. The viable Solutions are then sent back to the clients so the engineers express their preferences on each of them by grading each with a value ranging from 0 to 10. The grades are then sent back to the Coordinator who takes the winner solution as described above. The winner solution is then notified to all the clients and applied to the model.

4   Conclusions

In this paper we presented a system designed to support collaborative work in Civil Infrastructure projects that is able to assist in the detection and solving of semantic Errors and Conflicts. These semantic problems, which also involve geometric problems, are so common that they are normally accepted so long they can be in-field detected and corrected. However, this is not always the case and they may eventually lead to project delays and to overheads. Thus, it is important that the models are delivered free of problems as much as possible. A semantically perfect model without problems or ambiguities eases the automation of the tasks, which translates to a more efficient usage of resources. Conflicts are a special case of problem which are especially difficult to solve. Negotiating is the natural mechanism to reach an agreement on how to solve them. Our system provides a structure for this negotiation by means of suggesting alternatives and picking the preferred one among all the parties -the Profiles- involved in the conflict. The preferred alternative is is the one that maximizes de global welfare.

Acknowledgements. This work was supported by Norwegian Research Council, Industrial PhD scheme case no: 195940/I40 through Vianova Systems AS, Norway; the Spanish MICINN, Consolider Programme and Plan E funds, as well as European Commission FEDER funds, under Grants CSD2006-00046 and TIN2009-14475-C04-04. It was also partly supported by Development and Planning Department of the Ålborg University (AAU), the Vice-rectorate for Research of the Universitat de València (UV) under grant UV-INV-AE11-40990.62. Authors also want to thank Dr. Erik Kjems from the AAU for his help without which this paper would not be possible.

References

1. Anumba, C., Ugwu, O., Newnham, L., Thorpe, A.: Collaborative design of structures using intelligent agents. Automation in Construction 11, 89–103 (2002)
2. Bellifemine, F., Caire, G., Greenwood, D.: Developing Multi-agent Systems with JADE. John Wiley & Sons Ltd. (2007)
3. Chevaleyre, Y., Dunne, P.E., Endriss, U., Lang, J., Lemaitre, M., Maudet, N., Padget, J., Phelps, S., Rodriguez-Aguilar, J.A., Sousa, P.: Issues in multiagent resource allocation. Informatica 30, 3–31 (2006)
4. Eastman, C., Teicholz, P., Sacks, R., Liston, K.: BIM Handbook, a guide to Building Information Modeling for Owners, Managers, Designers, Engineers, and Contractors. John Wiley & Sons, Inc., New Jersey (2008)
5. Fitzpatrick, G.A.: The Locales Framework: Understanding and Designing for Cooperative Work. PhD thesis, University of Queensland (November 1998)
6. Kolbe, T.H., Gröger, G., Plümer, L.: Citygml: Interoperable access to 3d city models. In: Geo-information for Disaster Management, pp. 883–899 (2005)
7. O'Connor, M., Knublauch, H., Tu, S., Grosof, B., Dean, M., Grosso, W., Musen, M.: Supporting Rule System Interoperability on the Semantic Web with SWRL. In: Gil, Y., Motta, E., Benjamins, V.R., Musen, M.A. (eds.) ISWC 2005. LNCS, vol. 3729, pp. 974–986. Springer, Heidelberg (2005)
8. Peña-Mora, F., Wang, C.-Y.: Computer-supported collaborative negotiation methodology. Journal of Computing in Civil Engineering, 64–81 (April 1998)
9. Ren, Z., Anumba, C.J.: Multi-agent systems in construction–state of the art and prospects. Automation in Construction 13, 421–434 (2004)
10. Schnellenbach, M., Denk, H.: An agent-based virtual marketplace for aec-bidding. In: Proceedings of the 9th International EG-ICE Workshop Advances in Intelligent Computing in Engineering, Darmstadt, Germany, pp. 40–48 (2002)
11. Udeaja, C., Tah, J.: Agent-based material supply chain integration in construction. In: Perspectives on Innovation in Architecture, Engineering and Construction, CICE, pp. 377–388. Loughborough University (2001)
12. W3C OWL Working Group. OWL 2 web ontology language document overview. Technical report, W3C (October 2009)
13. Xue, X., Ji, Y., Li, L., Shen, Q.: Cognition driven framework for improving collaborative working in construction projects: Negotiation perspective. Journal of Business Economics and Management (2010)

Simulation and Analysis of Virtual Organizations of Agents

Elena García, Virginia Gallego, Sara Rodríguez, Carolina Zato, Juan F. de Paz, and Juan Manuel Corchado

Abstract. Nowadays there is a clear trend towards using methods and tools that can help to develop multiagent systems (MAS). Thanks to the contribution from agent based computing to the field of computer simulation mediated by ABS (Agent Based Simulation) is obtained benefits like methods for evaluation and visualization of multi agent systems or for training future users of a system. This study presents a multiagent based middleware for the agents behavior simulation. The main challenge of this work is the design and development of a new infrastructure that can act as a middleware to communicate the current technology in charge of the development of the multiagent system and the technology in charge of the simulation, visualization and analysis of the behavior of the agents. The proposed middleware infrastructure makes it possible to visualize the emergent agent behaviour and the entity agent in a 3D environment. It also allows to design multi-agent systems considering organizational aspects of agent societies.

Keywords: Multiagent systems, Simulation, JADE, Repast.

1 Introduction

The contribution from agent based computing to the field of computer simulation mediated by ABS (Agent Based Simulation) is a new paradigm for the simulation of complex systems that require a high level of interaction between the entities of the system. Possible benefits of agent based computing for computer simulation include methods for evaluation of multi agent systems or for training future users of a system [6]. Many new technical systems are distributed systems and involve complex interaction between humans and machines, which notably reduce their usability. The properties of ABS makes it especially suitable for simulating this

Elena García · Virginia Gallego · Sara Rodríguez · Carolina Zato · Juan F. de Paz
Juan Manuel Corchado
Computers and Automation Department, University of Salamanca, Salamanca, Spain
e-mail: {elegar,sandalia,srg,carol_zato,fcofds,corchado}@usal.es

J.M.C. Rodríguez et al. (Eds.): Trends in PAAMS, AISC 157, pp. 65–74.
springerlink.com   © Springer-Verlag Berlin Heidelberg 2012

kind of systems. The idea is to model the behaviour of the human users in terms of software agents. However, it is necessary to define new middleware solutions that allow the connection on ABS a simulation software.

This paper describes the results achieved towards a multiagent-based middleware for the agents' behavior simulation. The middleware, called MISIA (*Middleware Infrastructure to Simulate Intelligent Agents*), allows simulation, visualization and analysis of the agent' behavior. MISIA makes use of technologies for the development of multiagent systems known and widely used, and combines them so that it is possible to use their capabilities to build highly complex and dynamic systems.

The article is structured as follows: Section 2 makes a review of the most important requirements and the reasons that led to the realization of this research. Sections 3 introduces a description of the middleware specifically adapted to the simulation of virtual organizations within dynamic environments. Finally, some results conclusions are given in Sections 4 and 5.

2 Background and Requirements

Nowadays, MAS (Multi-Agent Systems) are widely used in various fields due to their inherent properties such as autonomy, local view, decentralization, coordination or cooperation among agents to solve a general problem. MAS based on organizational concepts coordinate and exchange services and information; they are capable of negotiating and coming to an agreement; and they can carry out other more complex social actions. At present, research focusing on the design of MAS from an organizational perspective seems to be gaining most ground. The emergent thought is that modeling the interactions in a MAS cannot be related exclusively to the actual agent and its communication capabilities; instead, organizational engineering is necessary as well. The concepts of rules, norms and institutions [8] and social structures [18] are rooted in the idea of needing a higher level of abstraction, independent from the agent, that explicitly defines the organizations in which the agents reside. MAS developers have focused their efforts on the organizational aspects of agent societies, using the concepts of organization, norms, roles, etc. Virtual organizations (VO) [9] are a means of understanding system models from a sociological perspective. VO have been usefully employed as a paradigm for developing agent systems [9]. One of the advantages of organizational development is that systems are modeled with a high level of abstraction, so the conceptual gap between real world and models is reduced. Also this kind of system offers facilities to implement open systems and heterogeneous member participation.

There are several different organizational approaches and platforms: JADE [17], S-Moise+ [13], RETSINA [11], Jack [12], EIDE [7], RICA-J [20], JASON [2], SIMBA [3], THOMAS [4]. However, the designers must implement all the features of simulation, if required; for example, in cases where it is necessary to take into account for both microscopic features, such as specifying the language between agents, as macroscopic characteristics that make up the target in a simulation.

On the other hand, thanks to the contribution from agent based computing to the field of computer simulation mediated by ABS is obtained benefits like methods for evaluation and visualization of multi agent systems or for training future users of the system [4][21]. Mainly there are two ways for visualizing multiagent systems simulation: the agents interaction protocol and the agent entity. In the former, it is visualized a sequence of messages between agents and the constraints on the content of those messages. On the other hand, the latter method visualizes the entity agent and its iteration with the environment. Most software programs, such as JADE platform [1][17] and Zeus toolkit [5], provide graphical tools that allow the visualization of the messages exchanged between agents. The toolkits MASON [14], Repast [15][19] [16]and Swarm [22] provide the visualization of the entity agent and its interaction with the environment. There are other works like Vizzari et al. [23] where is developed a framework supporting the development of MAS-based simulations based on the Multilayered Multiagent Situated System model provided with a 3D visualization.

It is necessary a software that allows to join characteristics of multi-agent systems and agent-based simulations systems. It is also fundamental in the field of ABS considering the agents from the viewpoint of cooperative, where becomes more important the purpose of the society and the rules and regulations that govern and control the behavior of its entities. It is interesting to have a platform of agents that allows to design multi-agent systems with the possibility of simulation and analysis of results, and moreover provides a module to define VO. The platform, therefore, must support frames coordination between agents, in addition to being able to dynamically adapt to changes in its structure, goals or interactions.

3   MISIA Adapted to VO

The platform that meets the above requirements is called MISIA (*Middleware Infrastructure To Simulate Intelligent Agents*) [10]. It is a platform that acts as middleware for simulation and visualization of multi-agent systems. It results from the union of two existing agent platforms: JADE, a platform widely used that comply with the FIPA standards, and Repast, used in the field of Agent Based Simulation. In addition to the union of both platforms, includes an additional module that will define Virtual Organizations of agents.

The main concept introduced in this study is the notion of time in JADE, which means that the events that occurred on this platform, such as messages between agents, will be uniquely sorted thanks to the global time provided by Repast. A schematic approach of the platform shown in the figure below:

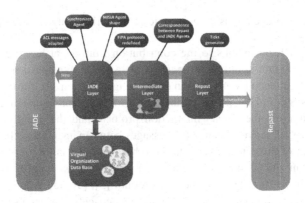

Fig. 1 MISIA Functional Approach

MISIA consists of four functional blocks:

1. *Jade Layer.* Its overall purpose is to adapt JADE to the simulation time. To this end, certain changes are required on the platform, such as: (i) designing a new template agent, which inherently manages time of the simulation; (ii) the adaptation of ACL messages containing current time information found in the simulation; (iii) a mechanism (synchronizer agent) to act as overall platform clock (in charge of the redistribution of-time events from Repast to JADE and of the notification to all JADE agents in terms of total ordering; (iv) or redefinition of the FIPA protocols provided by JADE which allow the programmer forgets tasks such as message handling and continues using these code templates in the new platform but, with the notion of time [10].

2. *Repast Layer.* It is a layer in direct contact with Repast. It is primarily responsible for the control of time of the platform: when an event goes by, it notifies the agent JADELlayer Synchronizer. In this module is where all agents represented in Repast inhabit .

3. *Intermediate Layer.* It is the union of the two adjacent layers. In addition to allowing the flow of information from one layer to another, manages the correspondence between agents of the type MISIA Agent of the JADE Layer and Layer Repast representatives. While an agent's reasoning and its communication properties will reside in the mold MISIA Agent, representation in the simulation is delegated to agents living in Repast Layer, so that an agent is divided in its logic and its figure.

4. *Virtual Organization Data Base.* It is a database connected to the platform that stores all the information about a virtual organization. This includes agents, work units, services, roles and societal norms.

4.1. *Agents* stores all the agents that are part of the virtual organization.

4.2 *Work Units* are the different groups that compose the VO. The platform is capable of managing any hierarchy of units no matter how many predecessors have a particular grouping.

4.3. *Services* acts as DF (Directory Facilitator) of a multi-agent platform with FIPA specification: it records all the services offered by agents.

4.4 *Roles* defines the different roles an agent can take. There is a direct relationship between the roles and services because for that an agent can play a certain role, it is necessary to offer services that role imposes by definition.

4.5 *Societal Norms* establishes rules that govern society. There are preset rules already on the platform, they are a series of rules that must be fulfilled by any VO, as is the fact that when an agent logs into a virtual organization must be clearly identified, or that is to say , must belong to a Work Unit, and must play a certain role; or the fact that only an agent who offers all services that a role requires, can adopt that role during his lifetime. The syntax of the rules that define the end user as well as compliance with them is delegated to the case study in particular. The objective of this idea is not to restrict the type of rules can be defined with a predefined syntax, leaving it open to increase the range of possibilities.

This module, in addition to serving as a database, ensures that these standards inherent to the platform are met at all times.

The access to the database of the virtual organization is done through an interface that provides the platform, that is, OVMisiaAgent agent services, manager of the same. These services allow all queries and modifications that obey the established rules. Following this scheme are clear potential bottlenecks: large numbers of requests for services of the same agent. In following sections, it proposes a scheme followed in a case study, which reduces this problem, through load sharing by means of permits, and intended to serve as a general model, regardless of the purpose of case study.

4 Experimental Results

It has been developed a case study to create a multiagent system aimed at facilitating the employment of people with disabilities, so it is possible to simulate the behavior of the agents in the work environment and observe the agents actions graphically. Every job is composed of a series of tasks. Agents representing the workers have to do them, and according to their capabilities, carry out the assignment with varying degrees of success. The main objective of this application is the search for the optimal arrangement for employees to achieve greater labor productivity in a given environment. And which would also serve as a prolepsis of the reality of situations, helping to improve the employability of people with disabilities.

The application is modeled as a multi-agent system where each element is represented as an agent: employees, jobs, architectural barriers, and so on. The office environment is simulated three-dimensional way with Unity 3D[1] engine, and allows to configure different architectural barriers, like a broken fire alarm, the door of an office too narrow, inclined ramps or a elevators broken.

---

[1] http://unity3d.com

The messages exchanged between different agents in the platform are translated into interactions, movement, or specific actions in the three-dimensional environment, according to the content of the message.

Therefore, the communication between the agent platform and Unity must be bidirectional: on the one hand, the creation of agents, or interactions between them must be notified from themulti-agent platform to simulation environment of Unity. On the other hand, when setting up an architectural barrier, it is necessary to notify of this change of state to the agent that represents it in MISIA. In the latter case it is necessary to send an event with source in Unity and destination in MISIA. This communication is implemented using sockets. Specifically, two TCP (Transmission Control Protocol) connections, one in each direction, following the client-server scheme, with both servers concurrently.

Because programming languages are different for each platform (MISIA uses *Java* and, in Unity, programming is done using scripts in *C#* and *Unity-JavaScript*), framework JNI (*Java Native Interface*) is used to enable this language translation. Because the volume of information exchanged between the two technologies are quite diverse (creation of agents, deletion of agents, interaction between different roles, etc.), it was necessary to design a communication protocol between MISIA and Unity that facilitates the interpretation of the information exchanged.

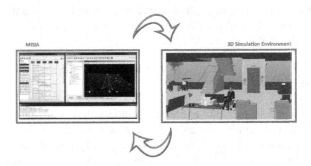

Fig. 1 Case Study: Accessibility in an Office Environment

The system is modeled as a VO that resembles reality, where the work units will be the departments that make up the office. It has three main Work Unit: the Quality Department, Human Resources Department and Production Department. Production department is subdivided according to the different occupations available in the virtual office, and the Quality Department has a sub-department where they are located all barriers. As shown in Figure 3, the topology followed for the design of the VO is a federation, as it is the best structure for the case study (a virtual office): each unit has an agent that represent it, which is the contact with other organizational units. This maintains a hierarchical structure by role of each employee.

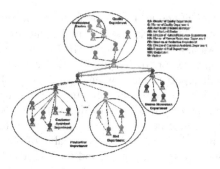

Fig. 2 Virtual Organization Structure

In the Production Department will be placing all worker agents and the agents representing jobs. To maintain a complete record of all activity performed by an worker, each time you perform a specific task, communication is maintained with their respective job (exchange of ACL messages according to the protocol FIPA more suitable in every situation). So so you can get information from the time the employee took to perform a given task and the start and end of the same.

4.1 Control Access to the VO

One of the most common problems that a dynamic VO always has to face is the behavior that it must adopt when an external agent appears. The agent's benevolence can be supossed but really, the cooperation is not guaranteed. It is needed to maintain an strict control of the agents that belong to the VO. The Human Resources Department of the organization's structure is in charge of this management that covers the entrance and exit of agents and the fulfillment of the norms defined by the end user.

In previous sections, the existence of the OVMisiaAgent has been mentioned. This agent is in charge of offering the services to interact with the database, which store the information related to the VO. Another one of the inherent norms of a VO included in MISIA is that an agent that does not own an specific authorization cannot access directly to the database. OVMisiaAgent owns this authorization from the beginning, reason why it can accede to the database and is enabled to offer all the management services related to database. This agent can give authorizations to other agents, so that they also can access to the database, without having to consume the services of OVMisiaAgent. This means that the charge can be distributed in a personalized way, which is the function carried out by the Human Resource Agents. These are agents playing a role that can own authorizations from OVMisiaAgent to access directly to the database. At any moment, indeed with simultaneous access, the consistency of the database is guaranteed. Moreover, this agents can define their own service interface allowing the fulfiment of the norms that have been previously defined in the case study.

In this particular case study, there are two different roles at the Department of Human Resources: Human Resources Worker (HR) and Human Resources Department Director (HB). HB keeps track of the amount of free jobs positions.

When an unemployed agent applies for a job, it communicates with a HR, and the HR first checks if he has the necessary skills (implements every service needed to adopt the role that it wants). Then, it asks the HB for any vacancy in the job that the unemployed is applying for. If everything is correct and the job can be assigned to him, HR will report to the new employee, giving him the information about his new job position and adding him into the VO. The agents that continuosly are interacting with the database are provided with the specific authorization: insertion of new agents (since in this case, the agents with the Employer role are not allowed to make modifications in the Agents table of the database, which is a norm defined previously in the case study and the RH agents must guard by its fulfillment all time), verification of its services, if those services already exist in the database or if all the services required by the role that wish to adopt are implemented. This way, the required cost to make requests to OVMisiaAgent is reduced to half regarding to the number of interchanged messages.

4.2 Learning in an VO

The main goal of Quality Department in this study case is to observe worker agents checking the time they spend to perform their tasks according to the architectural barriers that have affected them during the execution of that job. They are agents whose main goal is to learn depending on what they see. In this situation, in Quality Department, there are two main roles: Director of Quality Department (QB) and Worker of Quality Department(Q). The performance of this team is as follows: when a new employee joins the Virtual Organization, HR agent who gave him the job notifies the QB agent that a new employee exists and Quality Department must observe him. QB delegates this task to one of their employees, Q agents. At the beginning, there isn't any Q agent, but QB creates them depending on the number of employee agents in the Virtual Organization they have to study. Thus, this department adapts to the size of the Virtual Organization. Q agents tasks are to poll time to time to the employees who have been assigned to them, to know what tasks they are doing and what troubles they have when they execute their job, collecting details of every activity, and thus, future conclusions can be drawn by Quality Department, so they can readapt employees to different workstations if work efficiency can be improved with this decision.

This same scheme can be used in other kind of Virtual Organizations, when the goal of the study case is not as important as the capacity of this model to collect information in a decentralized way.

5 Conclusions and Future Works

The simulation of multiagent systems in several levels of details and the emergent behavior is fundamental for analyzing the systems processes. This study showed in detail the visualization and simulation infrastructure for developing the MAS behavior simulators. MISIA allows simulation, visualization and analysis of the behavior of agents. With the MAS behavior simulator it is possible to visualize the emergent phenomenon that arises from the agents' interactions.

Certain systems such as the case study presented in this article require well-defined topology. The aim of the simulations is to get more realistic, and this idea is linked to a great number of factors to consider. With the possibility of design of virtual organizations, we pretend reinforce the platform with this extra factors, approaching terms of social simulation. The agent representation is doing more photo-realistic: there is a 3D agent visualization in more levels of details showing the interaction them would make the system complete and realistic. Moreover, the interactivity with the user is allowed: the user can visualize the agent state and its simulation individually. The platform improve the interactivity by means of allowing the interaction of the specialists with the live execution besides the basic functionalities such as play, pause, stop and increase/decrease the speed, by means of putting some substances in the position and observing the emergent behavior. As a future line of work, we are considering the generation of reports about the information visualized during the simulation process in several levels of detail, which could increase the comprehension about the process. MISIA is the ideal framework for this purpose.

Acknowledgments. This work has been partially supported by the MICINN project TIN 2009-13839-C03-03.

References

[1] Bellifemine, F., Caire, G., Poggi, A., Rimassa, G.: Jade a white paper. EXP in Search of Innovation 3(3), 6–19 (2003)

[2] Bordini, R.H., Hübner, J.F., Wooldridge, M.: Programming Multi-Agent Systems in AgentSpeak Using Jason. John Wiley & Sons, Ltd. (2007)

[3] Carrascosa, C., Rebollo, M., Soler, J., Julian, V., Botti, V.: SIMBA Architecture for Social Real-Time Domains EUMAS 2003: 1st E. In: Workshop Multi-Agent Systems (2003)

[4] Carrascosa, C., Giret, A., Julian, V., Rebollo, M., Argente, E., Botti, V.: Service Oriented MAS: An open architecture. In: Decker, Sichman, Sierra, Castelfranchi (eds.) Proc. of 8th Int. Conf. on Autonomous Agents and Multiagent Systems (AAMAS 2009), Budapest, Hungary, May 10-15, pp. 1291–1292 (2009)

[5] Collis, J.C., Ndumu, D.T., Nwana, H.S., Lee, L.C.: The zeus agent building tool-kit. BT Technol. Journal 16(3) (1998)

[6] Davidsson, P.: Multi Agent Based Simulation: Beyond Social Simulation. In: Moss, S., Davidsson, P. (eds.) MABS 2000. LNCS (LNAI), vol. 1979, pp. 97–107. Springer, Heidelberg (2001)

[7] Esteva, M., Rodríguez-Aguilar, J.-A., Sierra, C., Garcia, P., Arcos, J.-L.: On the Formal Specification of Electronic Institutions. In: Sierra, C., Dignum, F. (eds.) AgentLink 2000. LNCS (LNAI), vol. 1991, pp. 126–147. Springer, Heidelberg (2001)

[8] Esteva, M.: Electronic Institutions: from specification to development. Ph.D. Thesis, Technical University of Catalonia (2003)

[9] Ferber, J., Gutknecht, O., Michel, F.: From Agents to Organizations: An Organizational View of Multi-agent Systems. In: Giorgini, P., Müller, J.P., Odell, J. (eds.) AOSE 2003. LNCS, vol. 2935, pp. 214–230. Springer, Heidelberg (2004)

[10]   García, E., Rodríguez, S., Martín, B., Zato, C., Pérez, B.: MISIA: Middleware Infra-
        structure to Simulate Intelligent Agents. In: Abraham, A., Corchado, J.M., González,
        S.R., De Paz Santana, J.F. (eds.) DCAI 2011. AISC, vol. 91, pp. 107–116. Springer,
        Heidelberg (2011) ISBN: 978-3-642-19933-2
[11]   Giampapa, J.A., Sycara, K.: Team-Oriented Agent Coordination in the RETSINA
        Multi-Agent System. Tech. Report CMU-RI-TR-02-34, Robotics Institute, Carnegie
        Mellon University, Presented at AAMAS 2002 (December 2002)
[12]   Howden, N., et al.: JACK intelligent agents-summary of an agent infrastructure. In:
        Proceedings of IEEE International Conference on Autonomous Agents, Montreal
        (2001)
[13]   Hübner, J.F., Sichman, J.S., Boissier, O.: S-Moise+:A Middleware for Developing
        Organized Multi-Agent Systems. In: Boissier, O., Padget, J., Dignum, V., Linde-
        mann, G., Matson, E., Ossowski, S., Sichman, J.S., Vázquez-Salceda, J. (eds.)
        ANIREM 2005 and OOOP 2005. LNCS (LNAI), vol. 3913, pp. 64–78. Springer,
        Heidelberg (2006)
[14]   Luke, S., Cioffi-Revilla, C., Panait, L., Mason, S.K.: A new multiagent simulation
        toolkit. In: Proceedings of the 2004 SwarmFest Workshop (2004)
[15]   North, M.J., Howe, T.R., Collier, N.T., Vos, J.R.: The repast symphony runtime
        system. In: Proceedings of the Agent 2005 Conference on Generative Social
        Processes, Models, and Mechanisms (2005)
[16]   North, M.J., Collier Nicholson, T., Vos Jerry, R.: Experiences Creating Three Im-
        plementations of the Repast Agent Modeling Toolkit. ACM Transactions on Model-
        ing and Computer Simulation 16(1), 1–25 (2006)
[17]   JADE, Java Agent Development Platform, http://JADE.tilab.com
[18]   Van Dyke Parunak, H., Odell, J.J.: Representing Social Structures in UML. In:
        Wooldridge, M.J., Weiß, G., Ciancarini, P. (eds.) AOSE 2001. LNCS, vol. 2222,
        pp. 1–16. Springer, Heidelberg (2002)
[19]   Repast, http://repast.sourceforge.net/repast_3/index.html
[20]   Serrano, J.M., Ossowski, S.: RICA-J -A Dialogue-Driven Software Framework for
        the Implementation of Multiagent Systems. In: JISBD Taller en Desarrollo de Sis-
        temas Multiagente (DESMA-2004), Málaga, pp. 48–61 (2004)
[21]   Shendarkar, A., Vasudevan, K., Lee, S., Son, Y.-J.: Crowd Simulation for Emergen-
        cy Response using BDI Agent based on Virtual Reality. In: Proceedings of the 2006
        Winter Simulation Conference, pp. 545–553 (2006)
[22]   Swarm, http://www.swarm.org
[23]   Vizzari, G., Pizzi, G., da Silva, F.S.C.: A framework for execution and visualization
        of situated agents based virtual environments. In: Workshop dagli Oggetti agli
        Agenti, pp. 22–25 (2007)
[24]   Wooldridge, M., Jennings, N.R.: Agent Theories, Architectures, and Languages:
        a Survey. In: Wooldridge, M., Jennings, N.R. (eds.) Intelligent Agents, pp. 1–22.
        Springer, Heidelberg (1995)
[25]   Zambonelli, F., Jennings, N.R., Wooldridge, M.: Developing Multiagent Systems:
        The Gaia Methodology. ACM Transactions on Software Engineering and Metho-
        dology 12, 317–370 (2003)

Using Simulation Based on Agents (ABS) and DES in Enterprise Integration Modelling Concepts

Pawel Pawlewski, Paul-Eric Dossou, and Paulina Golinska

Abstract. The aim of this paper is to share the dilemma about approach to simulation tool selection. The paper presents the results of the authors researches using methodologies of enterprises architectures (CIMOSA and GRAI) where agent approach is used to solve planning and managing problems. Processes are performed in enterprise manufacturing and supply chains. To verify new proposed solutions, simulation experiments are necessary. The problem is which simulation tool is appropriate to use for verification. Selected tools based on ABS and DES are presented. Some tools combining DES and ABS approaches are described. The process of choice and recommendation is also presented.

Keywords: Multi-agent systems, DES, simulation, Process modelling, Enterprise Architecture.

1 Introduction

The different economic and financial crises existing today have increased the necessity of enterprises to be prepared and well-organised. Enterprise modelling is one way for restructuring them in order to improve their performance and being more efficient. Three methodologies are mainly used for modelling enterprises: PERA, CIMOSA and GRAI. Enterprise modelling involves not only global enterprise performance improvement but also local improvements. Authors use in

---

Pawel Pawlewski · Paulina Golinska
Poznan University of Technology, ul.Strzelecka 11, 60-965 Poznań
e-mail: pawel.pawlewski@put.poznan.pl, paul-eric.dossou@icam.fr

Paul-Eric Dossou
ICAM, Site de Vendée, 28 Boulevard d'Angleterre, 85000 La Roche-Sur-Yon, France
e-mail: paulina.golinska@put.poznan.pl

J.M.C. Rodríguez et al. (Eds.): Trends in PAAMS, AISC 157, pp. 75–83.
springerlink.com   Springer-Verlag Berlin Heidelberg 2012

their researches CIMOSA and GRAI methodologies. For both multiagent systems was elaborated. In the case of CIMOSA it was agent system for planning process. For GRAI the research concerns a new tool based on agent system. The general structure of the tool is based on Case Based Reasoning (CBR). The CBR concepts are combined with Multi-agent systems for developing the new tool. Case Based Reasoning (CBR) remains widely used for the definition of the needs for the design and development of expert systems. The originality of this part is that it shows how the reasoning is combined with the theory of Multi-Agent systems and Artificial Intelligence and Metaphors of Mind.

In both case the verification of developed methods is necessary. Actually simulation is widely used and practically only one tool which can enable verification of complex systems.

The both presented cases are regarding manufacturing and supply chain processes. The logic of the planning and design of expert tool allows the use of agent technology but the simulation of the processes of manufacturing and supply chain is not so obvious. In authors' opinion in order to continue the further research, it is necessary to use simulators that take into account both the requirements of the agent approach and the requirements of classical DES (Discrete Event Systems). In this paper both approaches are discussed. The authors present available commercial simulation tools and define the requirements from point of view of the potential users- engineers dealing with operations management and supply chain processes.

The paper presents brief theoretical introduction to the studies (Section 2). Section 3 describes the management using MAS and model ling of manufacturing and supply chain processes. The overview of tools for Discrete Event Simulation DES is provided in Section 4. Discussion on the selected tools is presented in Chapter 5 and final conclusions are stated in Section 6.

2   VLPrograph and GraiMOD

CIMOSA is used for improving enterprises locally. The results of the previous studies are presented in paper [4]. These results consisted in planning the process taking place in an enterprise characterized by the manufacture of complex products (machine building). The idea is based on the so-called domains of the CIMOSA concept which has been used for the modelling. A planning process based on the multi-agent architecture called VLPRO-GRAPH. Agent-based system is defined in the present section as a multi-agent system that acts as a support tool and utilizes the databases of main system (ERP system). Multi-agent system is a collection of heterogeneous, encapsulated applications (agents) that participate in the decision making process [3]. The architecture of the tool (VLPRO-GRAPH – Very Long Process Graph) is based on the assumption that the system will support the MPS creation in ERP system and will be plugged in to ERP system database by for example java connector. This architecture was introduced in [5] and extended by a new agent, i.e. MR agent (movable resource agent). The task of this agent is integrated with the planning process which is described at three layers reflecting to [4]: A – the whole process perspective, the so-called whole process planning; B – the entity level where the whole process plan is divided into sub-plans which are executed by each sub- process and being transformed for individual production schedule at the domain level and where local re-planning activities takes place; C - domain sub-layer where production control activities are executed and information about disturbances is gathered and passed to upper levels.

GRAI Methodology is used for global performance improvements. GRAI Methodology is designed and defined for managing of this modelling. This method is used for example to choose and implement a computer tool (Supply Chain management and ERP) which meets the real market needs (globalisation, relocation, capacity to be proactive, cost optimisation, lead time, quality, flexibility, etc....). GRAIMOD is a tool being developed for supporting the methodology [1]. The general structure of the tool is based on Case Based Reasoning (CBR). According to CBR concepts the new case studied could be capitalized, but the parameters would also improve the reference model of the enterprise domain. Java is chosen for developing GRAIMOD. The Jade platform is being used in relation with FIPA-ACL for developing the different modules of GRAIMOD (GRAIQUAL for managing quality approach, GRAISUC for choosing and implementing an ERP or SCM tool in an enterprise and GRAIXPERT for managing reference models and rules used to improve enterprise performance).

The use of multi-agent systems will allow to facilitate the development of GRAIMOD. Some changes could be integrated according to the opinion of Jade specialists. Then CBR needs to be related to Multi-agents systems in order to satisfy user requirements. The reactive agents are not appropriate to our problem because they react only for the environment changes.

The global objective of this research is to be more efficient in the improvement of enterprises. The supply chain of each enterprise could be reorganized by using the concepts elaborated. The reorganization takes into account both the production typology and the supply chain in the modelling.

For each enterprise, the supply chain is decomposed into different parts (sourcing, procurement, purchasing, production, distribution, sales, transport and logistics management). For each part GRAIQUAL is used and a quality approach is defined in order to improve this part. Indeed, the optimizing of each part is coherent with the other parts.

For instance, the implementation of SQA in enterprise needs the use of knowledge relating to this enterprise domain, but the system will also evolve during this implementation. The system both provides the new case with data and takes into account the particularity of this new case. The multi-agent system defined is well-suited to this kind of work: use and capitalization of knowledge.

Multi-agent systems architecture also facilitates the communication between each different module of GRAIMOD by defining connections from each module to the others. The improvement of quality in the whole enterprise has a positive impact on cost and on delivery date.

3 Process Simulation – DES and ABS

For manufacturing and supply chain process simulation DES (Discrete-Event Simulation) has been mainstay for over 40 years. DES is useful for problems that consist of queuing simulations or complex network of queues, in which the processes can be well defined and their emphasis is on representing uncertainty through stochastic distributions [7]. Many of these applications occur in manufacturing, supply chain and service industries as well as queuing situations.

DES models are characterized by [7] process oriented approach (focus is on modelling the system in detail, not the entities). They are base on top down modelling approach and have one thread of control (centralised). They contains passive entities (i.e. something is done to the entities while they move through the system) and intelligence (e.g. decision making) is modelled as a part in the system. In DES queues are a key element; a flow of entities through a system is defined; macro behaviour is modelled and input distributions are often based on collect/measured (objective) data. These attributes describe manufacturing and supply chain processes too.

ABS (Agent Based Simulation) help to better understand real-world systems in which the representation or modelling of many individuals is important and for which the individuals have autonomous behaviours. ABS offers something novel, interesting and potentially highly applicable to manufacturing and supply chain. However, there is relatively little evidence that ABS is much used in the Operational Research community, there being few publications relating to its use in OR and OR-related simulation journal. Much greater volume of ABS papers is in journals from disciplines such as Computer Science, the Social Sciences and Economics.

Summarized ABS models are characterized by [7]:

-   Individual based (bottom up modelling approach); focus is on modelling the entities and interactions between them;
-   Bottom up modelling approach;
-   Each agent has its own thread of control (decentralised);
-   Active entities, i.e. the entities themselves can take on the initiative to do something; intelligence is represented within each individual entity;
-   No concept of queues;
-   No concept of flows; macro behaviour is not modelled, it emerges from the micro decisions of the individual agents;
-   Input distributions are often based on theories or subjective data;

These attributes doesn't describe manufacturing and supply chain processes but describe many aspect of management.

The emergence of ABS as a technique in Operational Research is timely. Globalised business is a highly complex management process, and making decisions in this environment is not well supported by the current set of tools, including DES [1].

4 Available Agent Systems

In table 1 is presented list of selected agent systems. This list includes systems which originate really from agent based approach.

Apart from presented list there are available many systems based on Java like: iGen, ICARO-T, JABM, JAMEL, JANUS, JAS, JASA , JCA-Sim, Madkit, Mason, Moduleco, Sugarscape, VSEit. There is a number of excellent academically developed tools, the commercially available software is limited to AnyLogic (but Anylogic origins are from DES so we classified it as DES system which included ABS approach), and all of these products expect knowledge of object oriented programming techniques and the modeller needs to be comfortable with Java. It is difficult to find an agent system which has possibilities to combine agent based and DES.

Table 1 List of selected available agent systems.

Name	Description	www
Altreva Adaptive Modeler	software application for creating market simulation models for price forecasting of real-world stocks and other securities.	www.altreva.com
AgentBuilder	an integrated software toolkit to quickly develop intelligent software agents and agent-based applications	www.agentbuilder.com
AOR Simulation	AB discrete event sim.; special extensions for modelling cognitive agents (with beliefs and speech-act-based information exchange communication).	oxygen.informatik.tu-cottbus.de/aor/
Ascape	General-purpose agent-based models.	ascape.sourceforge.net
Brahms	Multi-agent env. for sim. organizational processes	www.agentisolutions.com
Construct	Multi-agent model of group and organizational behavior.	www.casos.cs.cmu.edu/projects/construct/index.php
FAMOJA	Resource flow management, theoretical systems science, applied systems, environmental analysis	www.usf.uos.de/projects/famoja/
JADE	Distrib applications composed of autonomous entities	jade.tilab.com/
NetLogo	Social and natural sciences; Help beginning users get started authoring models	ccl.northwestern.edu/netlogo/

5 DES Systems on the Market

Table 2 presents the list of selected DES systems. This list includes systems which really originate from discrete events approach.

Table 2 List of selected DES systems available on market (O – Open Source, C – Commercial).

Name	Description	O/C	www
PowerDEVS	an integrated tool for hybrid systems modeling and simulation based on the DEVS formalism.	O	www.fceia.unr.edu.ar/lsd/powerdevs/index.html
SimPy	an open source process-oriented discrete event simulation package implemented in Python.	O	simpy.sourceforge.net/

Table 2 (*continued*)

Tortuga	an open source software framework for discrete-event simulation in Java.	O	www.ohloh.net/p/tortugades
Facsimile	discrete-event simulation/emulation library	O	www.facsim.org/
Galatea	the product of two lines of research: simulation languages based on Zeigler's theory of simulation and logic-based agents.	O	galatea.sourceforge.net
MASON	fast discrete-event multiagent simulation library core in Java	O	cs.gmu.edu/~eclab/projects/mason/
AnyLogic	graphical general purpose simulation tool which supports discrete event (process-centric), system dynamics and agent-based modeling approaches	C	www.xjtek.com/
Arena	simulation and automation software developed by Rockwell Automation. It uses the SIMAN procesor and simulation language.	C	www.arenasimulation.com
Enterprise Dynamics	simulation platform developed by INCONTROL Simulation Software. Features include drag-and-drop modeling and instant 2D and 3D Animation	C	www.incontrolsim.com
ExtendSim	general purpose simulation software package	C	www.extendsim.com
Flexsim	discrete event simulation software which includes the basic and three product lines: distributed simulation system (DS), container terminal library (CT) and Healthcare Simulation (HC)	C	www.flexsim.com
Witness	A discrete event simulation environment, with graphical 2D & 3D and scripting interfaces, for modelling processes and experimentation	C	www.lanner.com
Plant Simulation	by Siemens PLM Software enables the simulation and optimization of production systems and processes	C	www.plm.automation.siemens.com
ProModel	discrete event simulation tools	C	www.promodel.com
Simio	tool for rapid modeling of discrete-event systems to give rapidly an accurate 3D animated models.	C	www.si mio.com

Some systems offer possibilities to combine DES with ABS. the first is AnyLogic. AnyLogic supports agents in a continuous or discrete environment and also supports sophisticated animation capabilities to visualize agent behaviours. It contains a graphical modelling language and also allows the user to extend simulation models with Java code. The Java nature of AnyLogic allows model extensions via Java coding as well as the creation of Java applets which can be opened with any standard browser. The second system is Simio. The Simio framework is a graphical object-oriented modelling framework as opposed to simply a set of classes in an object-oriented programming language that are useful for simulation modelling. The graphical modelling framework of Simio fully supports the core principles of object oriented modelling without requiring programming skills to add new objects to the system. Simio framework is domain neutral, and allows objects to be built that support many different application areas. The Simio framework supports multiple modelling paradigms. The framework supports the modelling of both discrete and continuous systems, and supports an event, process, object, and agent modelling view. The third system is Flexsim. The Flexsim Simulation Software is a new generation of simulation software. The from the scratch own developed simulation kernel, the seamless integration of Microsoft C++ and the use of the newest OpenGL technology for unrivalled 3D animation in combination with the just as compact as practice-oriented library are the highlights of Flexsim. Flexsim is offered in following versions: GP (General Purpose Simulation), CT (Container Terminal Simulation), DS (Distributed Simulation) and HC (Healthcare Simulation). Flexsim DS is an object oriented simulation tool. Thereby is naturally well suited to ABS. It also possesses special features for modelling large volume systems, either through it's built in modelling constructs, or c++ for especially demanding agent based models.

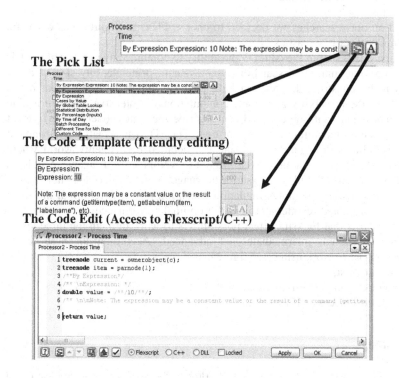

Fig. 1 Different level of user possibilities in Flexsim.

6 Conclusions

We identify two main barriers for ABS implementation in area of manufacturing and supply chain. These barriers are on the different levels:

- features of manufacturing and supply chain processes - queuing simulations or complex network of queues, in which the processes can be well defined and their emphasis is on representing uncertainty through stochastic distributions,

- all of ABS products expect knowledge of object oriented programming techniques and the modeller needs to be comfortable with Java. These are not skills that the average manager has developed during his career. For this reason, ABS remains the domain of a relatively few skilled experts and academic researchers.

The first challenge is therefore for the software development community, working in collaboration with current users from manufacturing and supply chain areas to establish how and where software can simplify the more technical aspects of ABS and reduces this barrier to entry. Reducing the amount of java code to be written is a must [7].

Based on our researches we decided to choose Flexsim (figure 1) as our main simulation tool for two reasons.

The first one is that in Flexsim DS the ability to combine any number of models together provides unlimited scalability, such that in principle, any size of agent model can be constructed. This can be especially important when agent based models have the possibility to become computationally intensive. Flexsim is able to represent agents with objects and can describe the state models of each agent object in it's own modelling language or c++. The 3D virtual reality environment of Flexsim allows the agents to operate in a detailed high fidelity world where geometry, shapes and motion exist. Rapid development of agent models is facilitated through the built in flexscript language engine which does not require compilation steps, and if more simulation execution power is required, the same code can be promoted seamlessly to c++ for optimal performance.

The second is that Flexsim offers possibilities to work through the three levels of users: occasional, intermediate and advanced. According to these levels Flexsim propose to work using (see figure 4):
- the pick list,
- the code template (user friendly),
the code edit (access to Flexsript/C++).

References

1. Dossou, P.-E., Mitchell, P., Pawlewski, P.: How to Successfully Combine Case Based Reasoning and Multi-Agent Systems for Supply Chain Improvement. In: Corchado, J.M., Pérez, J.B., Hallenborg, K., Golinska, P., Corchuelo, R. (eds.) Trends in PAAMS. AISC, vol. 90, pp. 75–82. Springer, Heidelberg (2011)
2. North, M.J., Macal, C.M.: Managing business complexity: Discovering strategic solutions with agent-based modeling and simulation. Oxford University Press, New York (2007)
3. Pechoucek, M., Říha, A., Vokrínek, J., Marík, V., Prazma, V.: ExPlanTech: Applying Multi-agent Systems in Production Planning. Production Planning and Control 3(3), 116–125 (2003)
4. Pawlewski, P.: Manufacturing Material Flow Analysis Based on Agent and Movable Resource Concept. In: Corchado, J.M., Pérez, J.B., Hallenborg, K., Golinska, P., Corchuelo, R. (eds.) Trends in PAAMS. AISC, vol. 90, pp. 67–74. Springer, Heidelberg (2011)

5. Pawlewski, P., Kawa, A.: Production Process Based on CIMOSA Modeling Approach and Software Agents. In: Demazeau, Y., Dignum, F., Corchado, J.M., Bajo, J., Corchuelo, R., Corchado, E., Fernández-Riverola, F., Julián, V.J., Pawlewski, P., Campbell, A. (eds.) Trends in PAAMS. AISC, vol. 71, pp. 233–240. Springer, Heidelberg (2010)
6. Sen, S., Weiss, G.: Learning in Multiagent Systems. In: Weiss, G. (ed.) Multiagent Systems: A Modern Approach to Distributed Artificial Intelligence, ch. 6, pp. 259–298. The MIT Press, Cambridge (1999)
7. Siebers, P.O., Macal, C.M., Garnett, J., Buxton, D., Pidd, M.: Discrete-Event Simulation is Dead, Long Live Agent-Based Simulation! Journal of Simulation 4(3), 204–210 (2010)
8. Wooldridge, M.: Intelligent Agents. In: Weiss, G. (ed.) Multiagent Systems:A Modern Approach to Distributed Artificial Intelligence, ch. 1, pp. 27–77. The MIT Press, Cambridge (1999)

A Genetic Algorithm-Based Heuristic for Part-Feeding Mobile Robot Scheduling Problem

Quang-Vinh Dang, Izabela Ewa Nielsen, and Grzegorz Bocewicz

Abstract. This present study deals with the problem of sequencing feeding tasks of a single mobile robot with manipulation arm which is able to provide parts or components for feeders of machines in a manufacturing cell. The mobile robot has to be scheduled in order to keep machines within the cell producing products without any shortage of parts. A method based on the characteristics of feeders and inspired by the (s, Q) inventory system, is thus applied to define time windows for feeding tasks of the robot. The performance criterion is to minimize total traveling time of the robot in a given planning horizon. A genetic algorithm-based heuristic is developed to find the near optimal solution for the problem. A case study is implemented at an impeller production line in a factory to demonstrate the result of the proposed approach.

Keywords: Scheduling, Mobile Robot, Genetic Algorithm, Part Feeding.

1 Introduction

Today's production systems range from fully automated to strictly manual. While the former is very efficient in high volumes but less flexible, the latter is reversed. Therefore, manufactures visualize the need for transformable production systems that combines the best of both worlds by using new assistive automation and mobile robots. A given problem is particularly considered for mobile robots with manipulation arms which will automate extended logistic tasks by not only transporting but also collecting containers of parts and emptying them into the place

Quang-Vinh Dang · Izabela Ewa Nielsen
Dept. of Mechanical and Manufacturing Engineering, Aalborg University, Denmark
e-mail: {vinhise, izabela}@m-tech.aau.dk

Grzegorz Bocewicz
Dept. of Computer Science and Management, Koszalin University of Technology, Poland
e-mail: bocewicz@ie.tu.koszalin.pl

J.M.C. Rodríguez et al. (Eds.): Trends in PAAMS, AISC 157, pp. 85–92.
springerlink.com   © Springer-Verlag Berlin Heidelberg 2012

needed. In that context mobile robots play the role of agents [12], attempting to reach their goals while following rules specific for a given production system. So, the considered systems are treated as multi-agent ones in which each robot can be seen as an autonomous object capable to undertake decisions about moving, feeding, emptying containers and completing operations, etc.

Feeding operation studied in this paper is a kind of extended logistic tasks. However, to utilize agents in an efficient manner requires the ability to properly schedule these feeding tasks. Hence, it is important to plan in which sequence agents' process feeding operations so that they could effectively work while satisfying a number of technological constraints.

Robot scheduling problem has attracted interest of researchers in recent decades. Crama and van de Klundert [1] considered the flow shop problem with one transporting robot and one type of product to find shortest cyclic schedule for the robot. Afterwards, they demonstrated that the sequence of activities whose execution produces one part yields optimal production rates for three-machine robotic flow shops [2]. Crame et al. [3] also presented a survey of cyclic robotic scheduling problem along with their existing solution approaches. Dror and Stulman [5] dealt with the problem of optimizing one-dimensional robot's service movements. Kats and Levner [7, 8] considered m-machine production line processing identical parts served by a mobile robot to find the minimum cycle time for 2-cyclic schedules. Maimon et al. [9] introduced a neural network method for a material-handling robot task-sequencing problem. Suárez and Rosell [11] built several strategies and simulation model to deal with the real case of feeding sequence selection in a manufacturing cell consisting of four identical machines. Most of the work and theory foundation considered approaches for scheduling robots which are usually inflexible, move only on fixed path and repeatedly perform a limited sequence of activities. There is still lack of approaches for scheduling mobile robots which are able to move around within a manufacturing cell to process extended logistic tasks with specific time windows and limitation in carrying capacity of mobile robots. Such a problem is modeled in several respects comparable to Asymmetric Traveling Salesman Problem (ATSP) in which the traveling time/cost might be different in two directions of a path. In addition, the ATSP belongs to NP-complete class [8] in which the required computational time exponentially grows with the size of the problem. Therefore, in this paper we focus on developing a computationally efficient heuristic, namely genetic algorithm-based heuristic for scheduling of mobile robot for feeding tasks which could be predetermined based on characteristics of feeders.

The remainder of this paper is organized as follows: in the next section, problem description is described while a genetic algorithm-based heuristic is presented in Section 3. A case study is investigated to demonstrate the result of the proposed algorithm in Section 4. Finally, conclusions are drawn in Section 5.

2   Problem Description

The work is developed for a real cell that produces parts for the pump manufacturing industry at a factory in Denmark. The manufacturing cell consists of a central warehouse, a single mobile robot (an agent), multiple machines which form production lines and feeders which are designed to automatically supply parts to these machines. The robot carries one or several small load carriers (SLCs) containing parts from the warehouse, moves to feeder locations, empty all parts inside SLCs to feeders, then returns to the warehouse so as to unload all empty SLCs and take filled SLCs. Each feeder (or task), which possesses its own characteristics such as maximum, minimum levels and consuming rate of parts, has to be served a number of times in order to keep manufacturing products, so the robot has a set of sub-tasks possessing time windows to carry out on each feeder during planning horizon. The manufacturing cell considered can be seen as an agent system that can be easily extended to the multi-agent one.

To enable the construction of a feeding schedule for the mobile robot, the following assumptions are made: a fully automatic mobile robot is considered in disturbance free environment; the robot can carry several SLCs at a time with limitation in payload; all tasks are periodic, independent, and assigned to the same robot; working time and traveling time of the robot between any two locations, in which either one of the locations could be a feeder or warehouse, are known; consuming rate of parts in a feeder is known; all feeders of machines have to be fed up to maximum level and the robot starts from the warehouse at the initial stage.

In order to accomplish all the movements with a smallest consumed amount of battery energy, the total traveling time of the robot is an important objective to be considered. Concerning computational time of the problem belonging to NP-complete class, it exponentially grows with the size of the problem (i.e. longer planning horizon, larger number of feeders). It is therefore necessary to develop a computationally effective algorithm that determines in which way the robot should supply the feeders with parts in order to minimize its total traveling time within the manufacturing cell while satisfying a number of technological constraints.

3 Genetic Algorithm-Based Heuristic

Among many meta-heuristics, genetic algorithm, a well-known method, is applied to develop a heuristic, shown in Figure 1, which is allowed to convert the aforementioned problem to the way that a near optimal solution could be found. The genetic algorithm-based heuristic consists of the following steps: (i) genetic representation and initialization, (ii) adjustment mechanism and fitness evaluation, (iii) selection, and (iv) crossover and mutation.

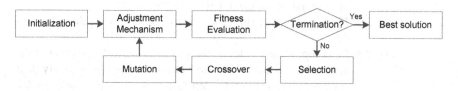

Fig. 1 Flow chart of the genetic algorithm-based heuristic

(i) Genetic Representation and Initialization

For the problem under consideration, a solution can be represented by a chromo-some as shown in Figure 2. Each gene in a chromosome consists of two parts. The first part is the index of a task (feeder) while the second one implies the index of sub-task of that task. The length of each chromosome is total number of sub-tasks of all tasks which the mobile robot has to perform during the planning horizon.

1,2	2,1	1,1	3,1	...	i,k	...	n,m

Fig. 2 Genetic representation

For initial generation, genes on a chromosome are randomly filled with sub-tasks of all tasks until the end of length of that chromosome.

(ii) Adjustment Mechanism and Fitness Evaluation

After initialization or crossover and mutation operations, chromosomes are ad-justed to be valid and then calculate their fitness values. A valid chromosome should satisfy two constraints about time windows of sub-tasks and capacity Q of the robot. For the first constraint, the start time of a sub-task of a task should be in-between release time and due time of that sub-task which could be determined by maximum level, minimum level and consuming rates of parts, while the second constraint requires the robot not to serve number of sub-tasks greater than number of SLCs it is carrying. An adjustment mechanism as below is applied to each chromosome in the initial generation or descendant so as to take these constraints into account.

Step 1: For each task, rearrange its sub-tasks in ascending order of their indices
Step 2: Rearrange considering time windows of all sub-tasks

> *Step 2.1:* Compute start time of the current sub-task. If start time of the current sub-task satisfies its time window then move to the next sub-task; otherwise go to step 2.2
> *Step 2.2:* Considering from the first sub-task to the current sub-task, make a list of candidates of sub-tasks whose release times are greater than that of the current one
> *Step 2.3:* Select randomly a candidate from the list
> *Step 2.4:* Insert the current sub-task to the position of the selected candi-date
> *Step 2.5:* Re-compute start times of the sub-tasks from the position of the selected candidate. If all start times of the sub-tasks satisfy their time windows then go to step 3; otherwise go back to Step 2.3. If none of candidate is selected then discard this chromosome, generate a new one instead and go back to Step 1

Step 3: Rearrange considering capacity of the robot and time windows

> *Step 3.1:* From the latest sub-task at the warehouse after every Q sub-tasks, add another sub-task at the warehouse of the robot.
>
> *Step 3.2:* Re-compute start times of the next Q sub-tasks from the newly added sub-task at the warehouse. If all start times of Q sub-tasks satisfy their time windows then go back to Step 3.1; otherwise go to Step 3.3
>
> *Step 3.3:* $Q = Q - 1$ and go back to Step 3.1. If $Q = 0$ then discard this chromosome, generate a new one instead and go back to Step 1. Note that the decrease of Q occurs only in the current loop, it will turn back to its predetermined value in the new loop.

Following the adjustment mechanism, the fitness evaluation will be taken place. The fitness value of a chromosome equals the total traveling time which the mobile robot moves from location of the first sub-task to location of the last one.

(iii) Selection

Various evolutionary methods can be applied to this problem. $(\mu + \lambda)$ selection is used to choose chromosomes for reproduction. Under this method, μ parents and λ offspring compete for survival and the μ best out of the offspring and old parents, in other words, the μ lowest in terms of the total traveling time, are selected as the parents of the next generation. Such selection mechanism guarantees that the best solutions up to now are always in the parent generation.

(iv) Crossover and Mutation

Crossover and mutation are main genetic operators. A crossover generates offspring by combining the information contained in the chromosomes of parents so that new chromosomes will have the good features of the parents' chromosomes. The Roulette-wheel selection is used in the algorithm, which probabilistically selects chromosomes of parents based on their fitness values (Goldberg, 1989). Genes on the selected chromosomes, which represent sub-tasks at the warehouse, are removed before recombination. An offspring then is generated with order crossover (OX) described as follow. Two cut points to be randomly chosen on the parent chromosomes. The string between these cut points in the first parent are first copied to the offspring. The remaining positions are filled by considering the sequence of genes in the second parent starting after the second cut point (when reaching to the end of chromosome, the sequence continues at position 1) [10]. The order crossover acts with probability P_c. After crossover, some offspring undergo mutation operator which is applied with probability P_m. The operation of mutation selects two positions within a chromosome at random and then inverts the substring between these two positions to produce heterogeneous chromosomes to avoid premature convergence of the algorithm.

4   Case Study

To examine performance of the proposed algorithm, a case study is investigated at the CR factory at Grundfos A/S. The chosen area for this case study is the CR 1-2-3 impeller production line which produces impellers for the CR products. The CR line consists of four feeders that have to be served by the mobile robot. Besides, different feeders are filled by different kinds of parts, namely back plates for feeder 1, vanes for feeder 2 and 3, front plates for feeder 4. On the CR line, a number of vanes are welded together with back and front plates to produce an impeller. Fig. 3 below particularly illustrates the aforementioned production area.

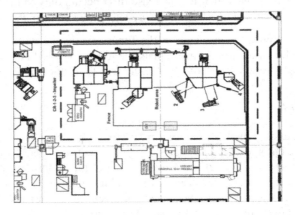

Fig. 3 CR 1-2-3 impeller production line

The maximum number of SLCs carried by the robot is 2. The average number of parts per SLC fed to feeder 1 or 4 is 125 (approximately 2 kg/SLC), while the average number of parts per SLC fed to feeder 2 or 3 is 1100 (approximately 1 kg/SLC). The maximum, minimum levels, consuming rate of parts and working time of robot at feeders are given in Table 1, while Tables 2 shows traveling time of robot from one location of feeder to another (feeder 0 means the warehouse).

Table 1 Maximum, minimum levels, consuming rate, and working time of robot at feeders

Feeder/Task	0	1	2	3	4
Maximum level (part)	-	250	2000	2000	250
Minimum level (part)	-	125	900	900	125
Consuming rate (sec/part)	-	4.5	1.5	1.5	4.5
Working time of robot (sec)	90	42	42	42	42

Table 2 Traveling time of robot from one location to another

Traveling time (sec)	0	1	2	3	4
0	-	49	44	43	38
1	49	-	58	45	58
2	46	58	-	35	48
3	42	43	35	-	47
4	44	56	47	46	-

The case study is investigated during 45 minutes because of battery limitation of the mobile robot. The proposed heuristic has been programmed in VB.NET and run on a PC that has Core i5 CPU, 2.67 GHz processor, and 4 GB RAM. A population size of 20 is used and probabilities of crossover and mutation are set to be 0.4 and 0.1, respectively. The termination is to stop at the generation of 100 or if no improvement is made after 50 generation. Figure 4 shows the convergence of the best solution of the proposed heuristic.

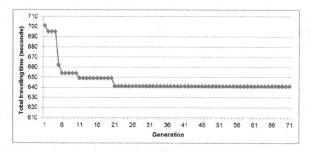

Fig. 4 Convergence of the best solution

The best solution obtained is given as: 0 - 4 - 1 - 0 - 4 - 1 - 0 - 4 - 3 - 0 - 1 - 1 - 0 - 4 - 2 - 0, with total traveling time being 641 seconds which makes up 23.7% of the planning horizon and the computational time of this case is less than a second. The proposed heuristic is significantly fast to find the near optimal solution in comparison with the optimal solution of 624 seconds and the computational time of 184 seconds which were obtained by the MIP model developed by Dang [4]. The Gantt chart of the best solution is shown in Fig. 5 below.

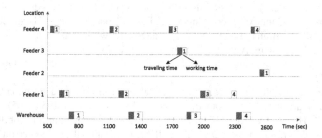

Fig. 5 Gantt chart for the best solution of the case study

5 Conclusions

In this paper, a new problem of scheduling a single mobile robot which performs part-feeding tasks is considered. To accomplish all tasks in the planning horizon within allowable limit of battery capacity, it is important to determine feeding sequences which minimize total traveling time of the mobile robot while taking into account a number of practical constraints. A genetic algorithm-based heuristic was developed to find near optimal solutions for the problem. The particular real case of an impeller production line was described to demonstrate the result of the proposed heuristic. The result showed that the proposed heuristic was significantly fast to obtain the near optimal solution. The solution is useful to managers for decision making at an operational level and the approach could be also applied in variety of tasks of not only mobile robots but also automatic guided vehicles.

The presented approach provides a solid framework that enables to model and evaluate scheduling tasks of multi-agent systems. For further research, re-scheduling mechanisms based on obtained schedules and feedback from the mobile robot fleet and shop floor should be developed to deal with real-time disturbances such as broken-down machines or unexpected shortage of parts of feeders.

Acknowledgments. "This work has partly been supported by the European Commission under grant agreement number FP7-260026-TAPAS".

References

[1] Crama, Y., van de Klundert, J.: Cyclic scheduling of identical parts in a robotic cell. Oper. Res. 45, 952–965 (1997)
[2] Crama, Y., van de Klundert, J.: Cyclic scheduling in 3-machine robotic flow shops. J. Sched. 2, 35–54 (1999)
[3] Crama, Y., Kats, V., van de Klundert, J., Levner, E.: Cyclic scheduling in robotic flow shops. Ann. Oper. Res. 96, 97–124 (2000)
[4] Dang, Q.V., Nielsen, I.E., Steger-Jensen, K.: Scheduling a single mobile robot for feeding tasks in a manufacturing cell. In: Proc. of Int. Conf. Adv. Prod. Manag. Syst., Norway (2011)
[5] Dror, M., Stulman, A.: Optimizing robot's service movement: a one dimensional case. Comput. Ind. Eng. 12, 39–46 (1987)
[6] Goldberg, D.E.: Genetic algorithms in search, optimization, and machine learning. Addison-Wesley, New York (1989)
[7] Kats, V., Levner, E.: Parametric algorithms for 2-cyclic robot scheduling with interval processing times. J. Sched. (2010), doi:10.1007/s10951-010-0166-0
[8] Kats, V., Levner, E.: A faster algorithm for 2-cyclic robotic scheduling with a fixed robot route and interval processing times. Eur. J. Oper. Res. 209, 51–56 (2011)
[9] Maimon, O., Braha, D., Seth, V.: A neural network approach for a robot task sequencing problem. Artif. Intell. Eng. 14, 175–189 (2000)
[10] Potvin, J.Y.: Genetic algorithms for traveling salesman problem. Ann. Oper. Res. 63, 339–370 (1996)
[11] Suárez, R., Rosell, J.: Feeding sequence selection in a manufacturing cell with four parallel machines. Robot Comput. Integrated Manuf. 21, 185–195 (2005)
[12] Shyu, J.-H., Liu, A., Kao-Shing, H.: A multi-agent architecture for mobile robot navigation control. In: Proceedings Tenth IEEE International Conference on Tools with Artificial Intelligence, pp. 50–57 (1998)

ACA Multiagent System for Satellite Image Classification

Moisés Espínola, José A. Piedra, Rosa Ayala, Luís Iribarne,
Saturnino Leguizamón, and Massimo Menenti

Abstract. In this paper, we present a multiagent system for satellite image classification. With this aim we will describe a new classification algorithm based on cellular automata called ACA (Algorithm based on Cellular Automata). This algorithm can be modeled by agents. Actually, there are different classification algorithms, such as minimum distance and parallelepiped classifiers, but none is fullreliable in terms of quality. One of the main advantages of ACA is to provide a mechanism which offers a hierarchical classification divided into levels of reliability with a final quality optimized through contextual techniques. Finally, we have developed a multiagent system which allows to classify satellite images in the SOLERES framework.

1 Introduction

The satellite image classification is one of the most important techniques used in remote sensing that helps on interpreting a great deal of information contained in the spectral bands [1], grouping together the image pixels in a finite number of classes, basing on the spectral values of the bands [2]. The information obtained by the satellite sensors as digital levels is changed into a categorical scale easy to interpret by the expert analyst. The resulting classified satellite image is essentially a thematic map of the original image and pixels belonging to the same class will

Moisés Espínola · José A. Piedra · Rosa Ayala · Luís Iribarne
Applied Computing Group, University of Almería, Spain
e-mail: {moises.espinola,jpiedra,rmayala,luis.iribarne}@ual.es

Saturnino Leguizamón
Regional Faculty of Mendoza, National Technnological University, Argentina
e-mail: saturnino.leguizamon@frm.utn.edu.ar

Massimo Menenti
Aerospace Engineering Optical and Laser Remote Sensing, TUDelft, Netherlands
e-mail: M.Menenti@tudelft.nl

J.M.C. Rodríguez et al. (Eds.): Trends in PAAMS, AISC 157, pp. 93–100.
springerlink.com   © Springer-Verlag Berlin Heidelberg 2012

have similar spectral characteristics [3]. The pixels are labelled as the same class most closely resembles digitally [4]. There are a lot of satellite image classification algorithms, and the use of a particular one depends of the analyst knowledge about the study zone [5].

The classification algorithms work really well in non-noisy satellite images and if the spectral properties of the pixels determine the classes sufficiently well. However, if the images are altered with a gaussian impulse-type noise or there are essential changes in the pixels properties, the resulting image may have lots of tiny areas (often a pixel) which are misclassified. To sort out this classification error, we can use contextual information taking into account not only the spectral values but also its surrounding pixels. There are several contextual classification algorithms which use average values or texture description to improve the spectral classification, but none is full-reliable in terms of quality. This paper studies the development of a new classification multiagents based on cellular automata that offers a hierarchical classification divided into levels of reliability and improves the final quality. The rest of the paper is structured as follows. Section 2 describes the use of cellular automata to classify satellite images (ACA). Section 3 shows the results and conclusions. Finally, in section 4, we finish the paper by exposing future work.

2 Classification with Cellular Automata

Cellular automata [6, 7] have been widely used for environmental simulations like simulating snow-cover dynamics [8], modelling vegetation systems dynamics [9], detecting vibrio cholerae by indirect measurement of climate and infectius disease [10], simulating forest fire spread for the prediction of disasters [11] [12] and simulating land use dynamics [13] [14].

Cellular automata have been also applied in image processing, like image enhancement (noise-reduction filters) and edges detection [15].

There are few works using cellular automata applied to satellite image classification [16], but these works do not use contextual techniques for getting a hierarchical classification. This paper presents a new methodology for implementing a satellite image supervised classification Algorithm based on Cellular Automata (ACA). It classifies the image pixels basing on both spectral and contextual information of each pixel, and consequently it improves the results obtained by other supervised classification algorithms in the literature (i.e., classical minimum distance algorithm is only based on spectral values without contextual information added).

2.1 General ACA Architecture

In order to create a multiagent system for satellite image classification based on cellular automata, we have implemented the architecture shown in Fig. 1.

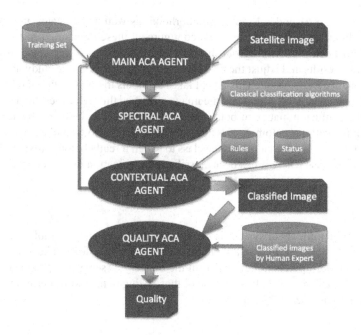

Fig. 1 The general ACA Multiagent Architecture.

The general ACA multiagents architecture is composed of 4 agents: the Main ACA Agent, the Spectral ACA Agent, the Contextual ACA Agent and the Quality ACA Agent:

- Main ACA Agent. The Main ACA Agent has two inputs: the original image that we want to classify and the training set obtained previously from pixels belonging to the classes selected for the cellular automata classification process. The Main ACA Agent manages the number of iterations of the cellular automata.
- Spectral ACA Agent. With inputs obtained by the Main ACA Agent, the Spectral ACA Agent is prepared to make a supervised classification, using results previously obtained by a modified classical supervised classification algorithm. The classical algorithms are minimum distance and parallelepiped supervised classifiers. Before carrying out the classification process, the user chooses the classical supervised algorithm on which he wishes the Spectral ACA Agent to be based.
- Contextual ACA Agent. In order to improve the results obtained through Spectral ACA Agent, the Contextual ACA Agent changes its behavior by using cellular automata, adding elements such as new states and rules to the supervised classification process. The user can also change the neighborhood (von Neumann,

Moore or extended Moore neighborhood) as well as the states and rules of the cellular automata. The user can configure all these inputs related to cellular automata before carrying out the classification process in order to customize the final results and adjust the cellular automata behavior to the study area.

- Quality ACA Agent. This Agent takes as inputs the image classified through the Contextual ACA Agent and the image classified through expert field work. It gets the confusion matrix of both images and it shows an index of the accuracy rate in the cellular automata classification process. It also provides a list of wrongly classified pixels that relates the class to which it really belongs (expert field work) to the class where it has been classified (classification of ACA Agents).

2.2 ACA Agents

The ACA multiagent system is shown in Table 1, 2 and 3. Table 1 executes the iterations of the cellular automata. In each iteration, we first make a spectral classification of all pixels not classified in the image yet (see Table 2) and subsequently we make a contextual classification of those being in the radius permitted belonging to their class (see Table 3).

Table 1 Main ACA agent.

Main ACA Agent $(E, numClasses, numIterations, threshold)$
Input:
$E = \{e_1, e_2, ..., e_n\}$: set of pixels to classify
$numClasses$: number of classes
$numIterations$: CA maximum iterations
$threshold$: threshold for class membership
Output:
$C = \{c_1, c_2, ..., c_k\}$: set of classes centers
$L = \{l(e)|e = 1, 2, ..., n\}$: classified pixels
01   **for** $i \leftarrow 0$ to $numIterations$ **do**
02   **foreach** $e_i \in E$ **do**
03   **if** $e_i.classificated \neq true$ **then**
04            $spectralClasses \leftarrow SpectralACAAgent(e_i, numClasses, threshold)$;
05   **if** $spectralClasses \neq \emptyset$ **then**
06              $finalClass \leftarrow ContextualACAAgent(e_i, spectralClasses)$;
07   **endif**
08   **endif**
09   **end**
10        $threshold \leftarrow threshold + incremental$;
11   **end**

Table 2 Spectral ACA agent.

Spectral ACA Agent ($e_i, numClasses, threshold$)
Input:
e_i: pixel to classify
numClasses: number of classes
threshold: threshold for class membership
Output:
spectralClasses: classes which may belong the pixel
01   *spectralClasses* $\leftarrow \emptyset$;
02   **for** $j \leftarrow 0$ to *numClasses* **do**
03   **if** *minDistance*$(e_i, class_j)\|j \in \{1..k\} \leq threshold$ **then**
04   *spectralClasses* $\leftarrow$ *spectralClasses* $\cup$ *class$_j$*;
05   **endif**
06   **end**

Table 3 Contextual ACA agent.

Contextual ACA Agent ($e_i, spectralClasses$)
Input:
e_i: pixel to classify
spectralClasses: classes which may belong the pixel
Output:
finalClass: final class of the pixel
01   **if** *size*$(spectralClasses) = 1$ **then**
02   **if** $\{spectralClasses\} \neq noiseClass$ **then**
03          $e_i.class \leftarrow \{spectralClasses\}$;
04          $e_i.quality \leftarrow numIteration$;
05   **if** *neighbourhoodClassesType*$(e_i) = 1$ **then**
06             $e_i.type \leftarrow focusPixel$;
07   **endif**
08   **if** *neighbourhoodClassesType*$(e_i) = 2$ **then**
09             $e_i.type \leftarrow edgePixel$;
10   **endif**
11   **endif**
12   **if** $\{spectralClasses\} = noiseClass$ **then**
13          $e_i.class \leftarrow bayesNeighbourhood()$;
14          $e_i.quality \leftarrow numIteration$;
15          $e_i.type \leftarrow noisePixel$;
16   **endif**
17   **endif**
18   **if** *size*$(spectralClasses) \neq 1$ **then**
19       $e_i.class \leftarrow bayesNeighbourhoodClass()$;
20       $e_i.quality \leftarrow numIteration$;
21       $e_i.type \leftarrow uncertainPixel$;
22   **endif**

3 Result and Conclusions

The satellite image classification algorithm based on ACA cellular automata has been implemented and experimented in the SOLERES framework. The experiments have been carried out on a multispectral Landsat image using a window size of 301x301 pixels (90601 pixels altogether). The spatial resolution of each pixel is 30x30m. The image corresponds with a region in the provinces of Almeria and Granada (Spain). This geographic region has a significant percentage of uncertain pixels due to the soil heterogeneity and the image has a minimum percentage of noisy pixels (1% of noise has been added artificially).

Fig. 2 shows the spectral ACA that is based on classical supervised algorithm (minimum distance) improved by means of cellular automata techniques thanks to the division of the classification process into several iterations. This division entails a hierarchical classification of different layers with a level of reliability each one. Therefore, the pixels classified in the first iterations are more reliable than the pixels classified in the following ones.

There are 77174 rightly classified pixels (85% rightly classified) in the classical minimum distance algorithm and 81599 rightly classified pixels (90% rightly classified) in the ACA minimum distance algorithm. Therefore, our algorithm improves the accuracy rate of the classical minimum distance algorithm around 5% by the confusion matrix.

In conclusion we can say that the results obtained in the satellite image classification using cellular automata are quite satisfying from several points of view.

Firstly, we improved the classification quality. We achieved this objective thanks to the contextual information provided by the neighborhood of the cellular automaton. Such process was improved because we used neighbors classified in previous iterations, and therefore, pixels closer to classes from a spectral point of view. This makes the uncertain and noisy pixels be classified with a higher degree of certainty. Our algorithm improves the accuracy rate of the classical minimum distance algorithm around 5%.

Secondly, with the ACA multiagent system we get a hierarchical classification based on layers of reliability, where each layer corresponds to a cellular automatons iteration. Throughout the execution of the ACA, the spectral radius belonging to the classes is increasing in each iteration; so, the first iterations provide maximum reliability layers since such layers have the pixels spectrally closest to the centers of the classes (some even correspond to the set of samples). While the cellular automaton is executing iterations, such spectral distance is increasing; so, the classified pixels are less reliable because they are further from the centre of their corresponding classes. Finally, the uncertain pixels are classified in the last iterations. They require contextual information to count the classes of the neighboring pixels (already classified in previous iterations, and therefore more reliable) and thus improve the final result.

Thirdly, the ACA multiagent system also offers spatial edge detection of classes in the satellite image itself, which can be rather useful in the subsequent

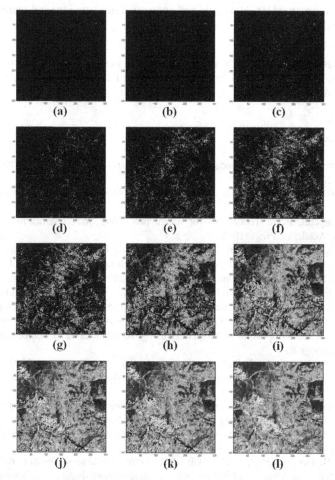

Fig. 2 (a) Iteration 1, (b) Iteration 3, (c) Iteration 5, (d) Iteration 7, (e) Iteration 9, (f) Iteration 11, (g) Iteration 13, (h) Iteration 15, (i) Iteration 17, (j) Iteration 20, (k) Iteration 50, (l) Iteration 100.

interpretation and analysis of the results obtained, as well as a list of uncertain and noisy pixels so that the experts can easily detect them.

4   Future Work

This section mentions the future objectives to optimize the presented work:

- Develop a fuzzy ACA multiagent system adding fuzzy rules and states to the cellular automata.
- Add a new level of classification to the ACA multiagent system: textural classification (based on textures). Thus, we would have two different levels of

classification: pixel level (spectral and contextual information) and regional level
(texture data).

Acknowledgements. This work has been supported by the EU (FEDER) and the Spanish
Ministry MICINN under grant of the TIN2010-15588 and TRA2009-0309 projects, and the
JUNTA DE ANDALUCIA under grant TIC-06114 project.

References

1. Chuvieco, E., Huete, A.: Fundamentals of satellite remote sensing. CRC Press, Boca
   Raton (2010)
2. Rees, W.G.: Physical principles of remote sensing, 2nd edn. Cambridge University Press
   (2001)
3. Schowengerdt, R.A.: Techniques for image processing and classification in remote
   sensing. Academic Press (1985)
4. Ayala, R., Becerra, A., Flores, I.M., Bienvenido, J.F., Diaz, J.R.: Evaluation of green-
   house covered extensions and required resources with satellite images and GIS. Almeria
   case. In: Second European Conference of the European Federation for Information Technology in Agriculture, Food and the Environment, Bonn, Germany, pp. 27–30 (1999)
5. Ayala, R., Menenti, M., Girolana, D.: Evaluation methodology for classification process
   of digital images. In: IEEE Int. Geoscience and Remote Sensing Symposium and the
   24th Canadian Symposium on Remote Sensing, IGARSS 2002, Toronto, Canada, pp.
   3363–3365 (2002)
6. Wolfram, S.: A new kind of science. Wolfram Media, Inc., Champaign (2002)
7. Kari, J.: Theory of cellular automata: a survey. Theoretical Computer Science 334, 3–33
   (2005)
8. Leguizamon, S.: Simulation of snow-cover dynamics using the cellular automata
   approach. In: 8th Symp. on High Mountain R. Sens. Cartography, pp. 87–91 (2005)
9. Balzter, H., Braun, P., Kuhler, W.: Cellular automata models for vegetation dynamics.
   Ecological Modelling 107, 113–125 (1998)
10. Lobitz, B., Beck, L., Huq, A., et al.: Climate and infectious disease: use of remote sensing
    for detection of Vibrio cholerae by indirect measurement. National Academic of Sci.
    USA 97(4), 1438–1443 (2000)
11. Karafyllidis, I., Thanailakis, A.: A model for predicting forest fire spreading using cellu-
    lar automata. Ecological Modelling 99, 87–97 (1997)
12. Muzy, A., Innocenti, E., Aiello, A., Santucci, J.F., Santonio, P.A., Hill, D.: Modelling and
    simulation of ecological propagation processes: application to fire spread. Environmental
    Modelling and Software 20, 827–842 (2005)
13. Leguizamon, S.: Modeling land features dynamics by using cellular automata tech-
    niques. In: ISPR Technical Comision, pp. 497–501 (2006)
14. Messina, J., Walsh, S.: Simulating land use and land cover dynamics in the ecuadorian
    Amazon through cellular automata approaches and an integrated GIS. In: Open Meeting
    of the Human Dimensions of Global Environmental Change Research Community in
    Rio de Janeiro, Brazil, pp. 6–8 (2001)
15. Popovici, A., Popovici, D.: Cellular automata in image processing. In: 15th Int. Symp.
    Mathematical Theory of Networks and Systems (2002)
16. Mojaradi, B., Lucas, C., Varshosaz, M.: Using learning cellular automata for post clas-
    sification satellite imagery. International Archives of Photogrammetry Remote Sensing
    and Spatial Information Sciences 35(4), 991–995 (2004)

Automatic Extraction of Geographic Locations on Articles of Digital Newspapers

Cesar García Gómez, Ana Flores Cuadrado, Jorge Díez Mínguez,
and Eduardo Villoslada de la Torre

Abstract. On this article, we present a model to make easier the reading of digital newspapers extracting the location of the news from the articles and showing the places associated with the news on a map. A module of supervised keyword-based extraction recognizes and classifies the geographical locations like named entities. The extraction results are improved using dictionaries or gazetteers (a list of named entities of the geographic area where the news are located). Thesauri are also used to check and complete the results, and for the named entities disambiguation. Finally, the model has been applied to *"El Norte de Castilla"*, a digital publication of Vallladolid, to validate and identify the tools and techniques with the best results.

1 Introduction

In the last years, the content of more and more digital publications (magazines, newspapers) has been published on the Internet. There are even publications which can be exclusively read on the Web. To compete on this area, digital publications, besides taking into account the news quality and its importance, must offer value added services to attract a greater number of readers, such as displaying the news on maps based on its geographic location.

Currently, there are applications that allow searches on a group of fixed newspapers based on keywords. E.g. users can search for news, activities or services related to a location of United Kingdom using UpMyStreet [24].

Cesar García Gómez · Ana Flores Cuadrado · Eduardo Villoslada de la Torre
Telefónica Investigación y Desarrollo,
Parque Tecnológico de Boecillo, Parcela 120, 47151, Boecillo, Valladolid, España
e-mail: {cesargg, anafc, evdlt}@tid.es

Jorge Díez Mínguez
Universidad de Valladolid, Edificio TIT,
Campus "Miguel Delibes" s/n, 47011, Valladolid, España
e-mail: jdiez25@yahoo.com

J.M.C. Rodríguez et al. (Eds.): Trends in PAAMS, AISC 157, pp. 101–108.
springerlink.com   © Springer-Verlag Berlin Heidelberg 2012

However, our goal is to create a more intuitive and visual way of reading digital newspapers, through a model which is able to extract the locations where the news happens or those ones whom the news refers to, in order that the news can be located on a map. The application of Geographical Location Extraction techniques [19, 18] on natural language documents has been used to built this model.

The geographical locations extraction is a subtask (called *Named Entities Recognition and Classification (NERC)* [18, 20, 25, 17, 16]) of a use case of the techniques to extract keywords and phrases (Keyword Extraction / Keyphrase Extraction) [13]. The named entities to be extracted, are usually classified in the following categories; Location names [LOC], Person names [PER], Organizations [ORG] and others [MISC].

The recognition and classification of named entities is based on the use of techniques of Machine Learning, Pattern Matching and Natural Language Processing. Many recognition and classification tools must be trained using a set of training documents manually classified. When the training is over, another set of articles manually classified is used to check the proposed tool, and the results are contrasted with the classification made by human raters. The success degree of these tools can be increased providing them a limited vocabulary (a keywords set (or its synonyms) extracted from a document). In addition, a model based on semantic techniques [26, 2] can be used to improve the process of geographical location, such as thesauri to confirm and complete the results and to resolve ambiguities. The semantic models set up the relation-ship between the news terms that are not addresses, (such as buildings, institutions,...) and the geographical units that represent them. The model is populated with characteristic keywords of an area and it forms part of the dictionary of streets, towns, organizations, institutions,..., which are readily available on Internet (street maps, *GIS (geographic information system) systems*).

Finally, this system has been implemented developing a Web application where the results of the extraction of locations of some sections of *"El Norte de Castilla"*, a real digital publication, can be viewed.

The article contains the following parts: Section 2 describes the system architecture and its components, Section 3 presents and compares the results obtained of the improvements application (NERC tools, thesauri...) on the system. Finally Section 4 presents the conclusions and the future lines.

2   System Architecture

The architecture and the workflow of system operations are shown below (Figure 1):

1. The system connects to the *URL (Uniform Resource Locator)* where there are various RSS feeds *(Really Simple Syndication)* with a summary of the news of the digital newspaper. The system reads the contents of the RSS feeds and extracts the complete URLs of the news.
2. The news are read one by one and written into a file, only containing the title and the news text. The HTML *(HyperText Markup Language)* tags and other unnecessary elements are removed.

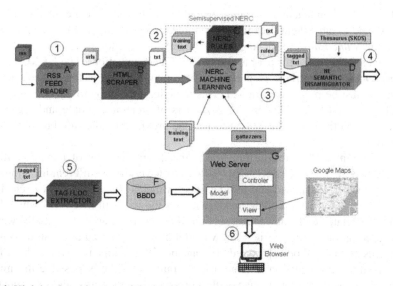

Fig. 1 High level architecture of the locations extraction system

3. Named entities of the news are labeled on the files by a semi-supervised NERC. The semi-supervised NERC is formed by the combination of a machine learning module, previously trained, and a rule-based NERC module
4. The disambiguation of named entities using thesauri and semantic techniques facilitates their identification on the files. The output files will contain the final labeled named entities.
5. In the next step, the named entities labels are extracted and stored on a database to facilitate subsequent consumption.
6. The Web application is in charge of showing a map with the possible locations of news where the users can interact with them.

2.1 Components

Each of the architecture components are briefly described on the next paragraphs.

- **RSS Feed Reader:** Reads the *XML (Extensible Markup Language)* / RSS files with the news summary whose URLs are stored on a configuration file, and extracts from them the full URL of each news.
- **HTML Scraper:** Performing HTML markup rules extracts the title and the body of the news from HTML files whose URLs have been obtained by the RSS Feed Reader. It removes the HTML marks and other unnecessary content (advertising, links to other sections of the newspaper, etc).
- **Semi-supervised NERC:** It has two modules: the **NERC module based on machine learning** and the **rule-based NERC**. The **rule-based NERC** module automatically tags the news not labeled manually. These tags are used to train the

NERC based on machine learning. The inputs of this module are the not labeled news and to label them uses the rules stored on a configuration file. The outputs are the *Name Entities* (NE) tagged on the articles depending on the type: [LOC] for locations, [PER] for people and [ORG] for organizations. The **NERC based on machine learning** is responsible of the final labeling of the named entities and it must be previously trained. This module, in addition to the output of the rule-based NERC, makes use of gazetteers of locations, people and organizations, providing a similar output to the one provided by the rule-based NERC, but improved.

- **NE Semantic Disambiguator:** Makes the disambiguation of named entities, classified by the semi-supervised NERC, that have several different meanings. It also checks if the named entities identified by the NERC module have been classified correctly. This module makes use of a thesaurus in SKOS [7] which has the NEs related to the newspaper section where the news appears. Initially, we've focused on the section of news of Valladolid, so we've used a thesaurus generated using the method [8] around the concept of the Wikipedia "*Valladolid*". The semantic disambiguation is done using a ranking of the N possible meanings that the named entity may have on the corresponding thesaurus and selecting the one that reaches the higher valuation as the meaning. To built the ranking, we've added a positively score when other words that maintains a semantic relationship (altLabel, broader and narrower) with the word we want to disambiguate appear on the same article. Unlike other solutions [3] that give the same score to all the words present in a term context, we give different weights to these words depending on their relation with the term to disambiguate

- **Tag and Localization Extractor:** It extracts the definitive named entities from the files with the news tagged by the previous modules. It stores the news on the database, linking each article with its set of named entities. It also makes use of simple heuristics to determine which is the location where the news are produced by applying precedence rules over the named entities of each article. These rules give priority to entities of LOC type over ORG and PER entities. Besides, if there are multiple LOC type entities, the less degree entity of the a political-administrative and geographic hierarchy is selected as location. (E.g. a street has preference over a city, if several entities are at the same level of the hierarchy, the one which appears more often, is chosen and in case of equality is selected the one which occurs above). A customized program has been created instead of using specific rule tools such as Drools [12], due to the rules simplicity.

- **Database:** Contains the tables where the news and the named entities are store in a structured way, facilitating the further consumption of the information by the Web tools.

- **Web Application:** Based on a *Model-View-Controller (MVC)* architecture, it is in charge of consulting the database and displaying the results properly. The user can view the located news of a date and for a particular newspaper section on a map and from there, he can access to the full news content on another section of the Website (figura 2).

Fig. 2 Web application screenshot over an article of a digital publication.

3   Experimentation and Validation

The evaluation purpose is to quantify the improvements introduced in the system with the use of the techniques presented in this paper. The technique defined at the IREX [11] and CoNLL [6] conferences has been used for the evaluation. This technique is based on the comparison of MAF *(micro-averaged f-measure)* (MAF or F1 = 2 * P * R / (P + R)). Where *precision (P)* is the percentage of correct named entities found and the *coverage or recall (R)* is the percentage of named entities present on the articles found.

For the NERC tools selection, it has been made a study of several of them, which have selected by their good results at conferences and congresses [18, 10, 5] and by holding a steep learning curve. The tools selected were LBJ NER Tagger 1.2 [14, 21], Stranford NER [22], Lingpipe [1], CAGEclass [4], DRAMNERI [23], LT-TTT2 [15], Freeling [9]. Several tests using different tools over the same set of test articles of *"El Norte de Castilla"* have been performed. For all the tools, we've performed some tests using local and general gazetteers and some tests without them. As a result of the experiment, we can conclude that the best results have been obtained using the LBJ NER Tagger 1.2 tool trained with the CoNLL2002 dataset and using local gazetteers (Figure 3).

Two techniques have been used to improve the results: The first one consists on including new locations (popular names of areas, neighborhoods or build-ings...), local business organizations (companies on the region, hotels and restau-rants, churches, museums, ..) on the used dictionaries, and increasing the number

NAMED ENTITIES	PRECISION	RECALL	F1
PER (Persons)	0.706	0.923	0.8
LOC (Locations)	0.72	0.72	0.72
ORG (Organizations)	0.739	0.58	0.65
GLOBAL	0.694	0.677	0.685

Fig. 3 LBJ NER. Training with CoNLL2002 and local gazetteers

of names of registered people. The other one is based on the thesauri use to make the disambiguation of the named entities and to confirm the classification made by the NERC tool. The thesauri are also used to help on the selection of the locations shown on the Web. The Figure 4 shows the improvement on the results made using these techniques.

NAMED ENTITIES	PRECISION	RECALL	F1
PER (Persons)	0.882	1	0.937
LOC (Locations)	0.864	0.704	0.776
ORG (Organizations)	0.882	0.833	0.857
GLOBAL	0.877	0.821	0.848

Fig. 4 LBJ NER. Training with CoNLL2002, improved with local gazetteers and thesauri use

The improvement on the results is evident. The global F1 parameter (related with all named entities) improves about 20% compared to the value obtained before introducing the improvements. Moreover, in the geographical locations (LOC entities) case, the F1 improvement is over 7%. The improvements got in other types of entities are even greater, due to the use of better local dictionaries. If we add to the dictionaries the records of most companies and organizations, which belong not only to the province but also to the autonomous region, the F1 of organizations will be increased around 31%. If we increase the number of names of people, the F1 will be improved on 17%.

Most failures happen on the entity classification, when a entity is classified like a wrong type. E.g. *José Zorrilla stadium* should have been labeled like a ORG type, but it was classified like a person [PER], due to this combination of words does not often appear on the training set and the Wikipedia reinforces this term like a person concept, since the principal meaning of *José Zorrilla* on the Wikipedia is related with the famous writer born in *Valladolid*.

In conclusion, the results of NERC tools can be improved by optimizing the local dictionaries, adding to them entities directly related to the articles analyzed scope and using one or more thesauri related to the geographic area (city, province, region) where the most articles of the digital publication are mainly located.

4 Conclusions and Future Lines

In this paper, we've presented a system and a Web application that offer a more intuitive and visual way to access and read of the news on digital newspapers, applying location extraction techniques. The location information extracted is used to facilitate the reading of online newspapers, identifying visually the news location extracted from the text and showing the places associated with the news on a map.

The use of NERC tools to extract locations on delimited domains provides significant improvements on the performance of some existing NERC systems. Several of these tools improve their performance completing their gazetteers with local named entities. So, our scientific contribution is developing a system to prove that the success rates of these tools can be substantially improved applying two techniques: adding to the dictionaries and gazetteers the named entities related to the area covered by the publication and using semantic techniques such as thesauri to check the results got by the NERC tool and to resolve the named entity ambiguities

There are several areas for future evolution of the system. For example, it could be incorporated new rules which do not limit the identification to independent entities to select the final location of the news and allow to identify the routes or neighborhoods. Likewise, it could be included features that allow users to search for news happened in a date range, or in a given location or even in a geographic area selected on the map.

Another possible future line would be the thesauri generation based on the different categories (sports, economy, society,...) and the sections of the publication to analyze. The thesauri could be selected automatically depending on the section of the newspaper containing the news that we want to locate geographically. Besides, to increase the thesaurus reliability, when it is generated, it would be interesting to integrate other data sources, not just the Wikipedia. Finally regular updates of gazetteer used by NERC tools should be considered to maintain a high success rate.

References

1. Baldwin, B., Carpenter, B.: LingPipe, http://www.alias-i.com/lingpipe/
2. Blázquez, L.M.V., Pascual, A.F.R., Ángel, M., Poveda, B.: Ingeniería ontológica: El camino hacia la mejora del acceso a la información geográfica en el entorno web. In: Subdirección General de Aplicaciones Geográficas del Instituto Geográfico Nacional. Avances En Las Infraestructuras De Datos Espaciales, p. 95 (2006)
3. Brugmann, H., Malaisé, V., Gazendam, L.: Disambiguating automatic semantic annotation based on a thesaurus structure. In: Proc. 14e Conference Sur le Traitement Automatique des Langues Naturelles, TALN 2007 (2007)
4. CAGEclass, http://cageclass.sourceforge.net/ (last visit January 2011)
5. Chinchor, N.: Overview of MUC-7/MET-2. In: Proc. Message Understanding Conference, MUC-7 (1999)
6. CoNLL-2011, http://www.clips.ua.ac.be/conll/ (last visit March 2011)
7. Drools: The Business Object Integration Platform, http://www.jboss.org/drools

8. Flores Cuadrado, A., Villoslada de la Torre, E., Peláez Gutiérrez, A.: Generación de Tesauros basado en Media Wiki. Actas de los Talleres de las Jornadas de Ingeniería del Software y Bases de Datos 3(6) (2009)
9. FreeLing Home Page, http://nlp.lsi.upc.edu/freeling/ (last visit April 2011)
10. Grishman, R., Sundheim, B.: Message Understanding Conference-6: A Brief History. In: Proc. 16th Conference on Computational Linguistics, USA, vol. 1, pp. 466–471 (1996)
11. IREX: Information Retrieval and Extraction Exercise, http://nlp.cs.nyu.edu/irex/
12. Isaac, A., Summers, E.: SKOS: Simple Knowledge Organization System primer (2008), http://www.w3.org/TR/skos-primer (last visit March 2011)
13. Keyphrase Extraction Algorithm. Technical Report. Computer Science Department, University of Waikato. Hamilton, New Zealand, http://www.nzdl.org/Kea/index.html
14. Learning Based Java. Cognitive Computation Group. Universidad de Illinois, EEUU, http://cogcomp.cs.illinois.edu/page/software_view/11 (last visit April 2011)
15. LT-TTT2. Language Technology-Text Tokenisation Tool, http://www.ltg.ed.ac.uk/software/lt-ttt2 (last visit March 2011)
16. Mansouri, A., Affendey, L.S., Mamat, A.: Named Entity Recognition Approaches. International Journal of Computer Science and Network Security 8, 339–344 (2008)
17. Marrero, M., Sánchez-Cuadrado, S., Lara, J.M., Andreadakis, G.: Evaluation of named entity extraction systems. In: Conference on Intelligent Text Processing and Computational Linguistics (CICLing 2009), pp. 47–58 (2009)
18. Nadeau, D., Sekine, S.: A survey of named entity recognition and classification. Linguisticae Investigationes 30, 3–26 (2007)
19. Ortega, J.M.P., Cumbreras, M.A.G., Vega, M.G., López, L.A.U.: Sistemas de Recuperación de Información Geográfica multilinges en CLEF. Procesamiento Del Lenguaje Natural 40, 129–136 (2008)
20. Ortega, J.M.P., Ráez, A.M., Santiago, F.M., López, L.A.U.: Geo-NER: un reconocedor de entidades geográficas para inglés basado en GeoNames y Wikipedia. Procesamiento Del Lenguaje Natural 43, 33–40 (2009)
21. Ratinov, L.: Design Challenges and Misconceptions in Named Entity Recognition, http://cogcomp.cs.illinois.edu/page/publication_view/199 (last visit April 2011)
22. Stanford Named Entity Recognizer. The Stanford Natural Language Processing Group, http://nlp.stanford.edu/software/CRF-NER.shtml (last visit April 2011)
23. Toral, A.: DRAMNERI: a free knowledge based tool to named entity recognition. In: Proc. 1st Free Software Technologies Conference, La Coruña, España, pp. 27–31 (2005)
24. UpMyStreet, http://www.upmystreet.com/ (last visit April 2011)
25. Vargas, J.D.: Reconocimiento de Entidades Nombradas en Textos no Estructurados. Technical Report. Universidad Nacional de Colombia (2008)
26. Zapater, S., Javier, J.: Ontologías para servicios web semánticos de información de tráfico. Revista digital Dialnet. Lectura en la Universitat de Valencia en 2006 (2006), http://dialnet.unirioja.es/servlet/tesis?codigo=7157 (last visit March 2011)

An Experiment to Test URL Features for Web Page Classification*

Inma Hernández, Carlos R. Rivero, David Ruiz, and José Luis Arjona

Abstract. Web page classification has been extensively researched, using different types of features that are extracted either from the page content, the page structure or from other pages that link to that page. Using features from the page itself implies having to download it before its classification. We present an experiment to proof that URL tokens contain information enough to extract features to classify web pages. A classifier based on these features is able to classify a web page without having to download it previously, avoiding unnecessary downloads.

1 Introduction

Web page classification has been extensively researched, using different types of features. In this context, features can be extracted from different sources, including: the web page content [11], [13], [14], like the bag of words the page contains or the number of images in it; the page structure [1], [2], [5], [15], like the distance between its different components; or from other pages that link to the page [7], [8], like the words in the anchor text.

Our hypothesis is that, in many web sites (excluding URL-friendly sites), the set of tokens extracted from a page URL contains enough information to classify that page. That is, we can calculate some features based exclusively on the information contained in those tokens, and those features are usable to build a web page classifier.

Inma Hernández · Carlos R. Rivero · David Ruiz
University of Seville
e-mail: {inmahernandez, carlosrivero, druiz}@us.es

José Luis Arjona
University of Huelva
e-mail: jlarjona@gmail.com

*Supported by the European Commission (FEDER), the Spanish and the Andalusian R&D&I programmes (grants TIN2007-64119, P07-TIC-2602, P08-TIC-4100, TIN2008-04718-E, TIN2010-21744, TIN2010-09809-E, TIN2010-10811-E, and TIN2010-09988-E).

J.M.C. Rodríguez et al. (Eds.): Trends in PAAMS, AISC 157, pp. 109–116.
springerlink.com　　　　　© Springer-Verlag Berlin Heidelberg 2012

We have observed that in those sites, for all URLs there are mainly two classes of tokens: on one hand, we have tokens whose appearance in a URL depends on the request made by the user to obtain that page. For example, URL `http://scholar.google.com/scholar?q=java` is obtained after issuing a query to Google Scholar site, and token "java" depends on the particular query. On the other hand, there are tokens that do not depend on the user request; instead, they are always part of the URL for each kind of URLs. In the former example, the token "scholar" is always included in any URL obtained after issuing a query in this site, in the same position.

Considering exclusively the latter type of tokens, we are able to build URL prototypes, that is, strings that represent a collection of URLs with a similar format. Each URL prototype has a different format, and usually, points to a different type of page, hence if we build a collection of prototypes for the site, each one representing a different type of pages, we are able classify any given page, by comparing its URL with the different prototypes, and finding the best match.

In this paper, we propose an experiment to calculate features for tokens in URLs to test our previous hypothesis, and confirm our observation. We define features for tokens that distinguish between these two different classes of tokens, and we perform an statistical analysis to test whether these two classes of tokens are distinguishable in every web site. Once we justify this fact, we can use these features to build the former URL prototypes, and use them to classify web pages.

In comparison with the other types of features that can be used to classify web pages, token features can be calculated without having to download the web page, just by analysing its URL, which avoids unnecessary downloads.

The rest of the article is structured as follows. Section 2 describes the related work; Section 3 presents the experiment performed to justify our hypothesis; finally, Section 4 lists some of the conclusions drawn from the research and concludes the article.

2 Related Work

We have identified some proposals in the area of web page classification using features extracted from URLs. Some of them are supervised, like [12], [16], [4], [3], while [6] and [15] are not supervised.

Kan et. al. [12] presented one of the first approaches to web page topic classification using only URLs. Their proposal consisted on tokenising URLs into tokens, and then extracting features from those, like the words they contain, their type or length, amongst others. These features are used to build a Maximum Entropy classification model. They work with a subset of the URLs to build a classification model, so they perform a reduced crawling to obtain their training set. However, this is a supervised proposal that requires a labeled set of training URLs. Furthermore, they aim at classifying pages belonging to more than one site, and relies on words being human-understandable, which is not always true.

Vidal et. al [15] propose a technique to analyse a single site, and automatically find pages that are similar to an example page that is given. To achieve so, the site is mapped, and URL patterns are generated for those pages that lead (directly or eventually) to those pages. To detect similarity between pages, the Tree Edit Distance is used. To build the training set, they have to previously crawl the entire site, download each page and then process them, which takes a significant amount of time. Also, it does not classify a page according to its content, but to its structural similarity to the given page. That means that pages containing information about different topics may be classified as the same class.

Zhu et. al. [16] propose a link classifier, instead of a web page classifier, although we include it in this framework due to their analysis of link features. They aim at classifying links according to their function inside the site. They propose a taxonomy of predefined link classes, depending on the function that each link performs in the page, namely: navigating, indexing, citing, recommending and advertising. They analyse links to extract visual, content and structural features, amongst others, and they build two types of supervised classifiers: SVM and decision trees. It is supervised proposal in which the classes are predefined in a rigid taxonomy. Furthermore, their goal is not directly related to information extraction, like ours, so they are not topic oriented; instead, their classifier may be used for user link recommendation.

Baykan et. al. [3] proposed a classifier that is similar to [12]. They tokenise URLs as well to extract features, but they apply different supervised classification algorithms, like SVM, Naïve-Bayes or Maximum Entropy, and compare the results. These algorithms need to be fed a list of words indicative for every topic that is to be classified. On a previous work [4], the authors used the same idea, but this time in order to classify pages according to their language, instead of their topic. Just like [12], it is a supervised technique that classifies pages from different sites.

Finally, Blanco et. al. [6] consider that every site is created by populating HTML templates with data from a database. Their goal is to cluster web pages so that each cluster contains pages following a certain template. They observed that URLs generated from the same template have a similar pattern, just like pages generated from the same template contain similar terms, so they proposed an algorithm for unsupervised classification that combines web page contents and its URL as features, by means of the minimum description length method (MDL). They require a large training set, so they crawl the entire site in their experiments. Furthermore, to improve the classification efficiency, features from the page itself are included in addition to the link-based features, which means that it must be downloaded previously.

In contrast to the former, our proposal is not supervised, and it does not require to crawl the whole site to build the classification model, as in [15] or [6]. We use a small subset of pages and URLs from the site, and we apply a statistical technique to extract classification features from those URLs.

3 Experiment

Next, we present the experiment to justify our hypothesis. First, we describe the process to obtain a set of tokens from a given web site and calculate features for those tokens. Then, we show the results of applying the former process to five academical sites, and we analyse those results.

3.1 Features Calculation

To calculate the feature value of each token in a URL, we consider not only the token, but the whole sequence of tokens from the beginning of the URL up to the token itself, which we call the token prefix. For each token in a URL, we define its feature value as the probability of the token prefix appearing in other pages from the same web site. Following the frequentist approach, we estimate that probability using the relative frequency of appearance of the token prefix. Therefore, we need to obtain a sample of pages, extract the URLs of all the links inside them, and finally calculate the relative frequency of their token prefixes. From here onwards, we will represent the feature value of a token X by F_X.

To obtain the sample, we perform a lightweight crawling over each site, to retrieve a representative sample of its pages. In this paper, we focus on dynamic pages, that is, pages that are behind a form that must be filled in and submitted. The result of those submissions is usually a hub, i.e., a list of results to the query, including a brief description of the result, along with a link to another page with the extended information. Therefore, we should fill in the forms with values such that the resulting query yields as many results as possible. In that case, the hub obtained in response contains a large number of links, and our sample is more representative.

From those hubs, we extract all links, and we decompose them into tokens. We use a tokeniser based on the standard for URIs defined in RFC 3986, according to which a URI is composed of a protocol (e.g., http, ftp or https); an authority, also known as domain name (e.g., google.scholar.com); a path, which is a sequence of segments separated by a slash charater ('/'); optionally followed by an interrogation mark ('?') and a query string, which is a sequence of parameters separated by the ampersand character ('&'), being each parameter of the form name '=' value.

Then, we calculate for each token, its feature value F_X, estimating the probability of the token prefix appearing in other hubs from the same site, by means of its relative frequency.

3.2 Experimental Results

We have chosen five academical sites: Google Scholar, Arxiv, Microsoft Academic Search, TDG Scholar and DBLP. For each of them, we performed the lightweight crawling to obtain a test set of hubs. In order for this set to be representative from the site, we chose a sample size that was sufficiently large, retrieving 100 hubs per site. Then, we extracted all links in those hubs and tokenised them, obtaining the test sets shown in Figure 1.

For each token, we calculate their feature value by estimating its probability, as described previously. As an example, in Figure 2 we show the feature values for some of the URLs extracted from TDG Scholar. We graphically represent URLs, tokens

Site	# Hubs	# Links	# Tokens
Arxiv	100	33748	121362
DBLP	100	30749	157295
Google Scholar	100	6247	38375
Ms Academic Search	100	9588	107888
TDG Scholar	100	11055	91493

Fig. 1 Experimental test sets

and their features using a tree-like structure, in which each node represents a different token, and a token t_1 is a son of another token t_2 when t_2 follows t_1 in a given URL. Every token is assigned a label, which indicates its node name, its feature value, and the token text. Each path from the tree root to a leaf represents a single URL.

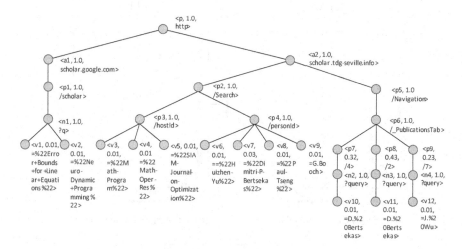

Fig. 2 Links extracted from TDG Scholar

In order to get an idea of how the feature values are distributed, we present some descriptive statistics in Figure 3. For each site, we calculate its mean, standard deviation, minimum and maximum values and quartiles.

Variable	N	Mean	Std. Dev	Min	Max	Percentiles		
						25th	50th	75th
Google_Fx	38375	0.04	0.13	0.01	1.00	0.01	0.01	0.02
DBLP_Fx	157295	0.02	0.07	0.01	1.00	0.01	0.01	0.01
MsAcademic_Fx	107888	0.02	0.07	0.01	1.00	0.01	0.01	0.01
TDG_Fx	91493	0.02	0.07	0.01	1.00	0.01	0.01	0.02
Arxiv_Fx	121362	0.02	0.04	0.01	1.00	0.01	0.01	0.02

Fig. 3 Descriptive statistics for the experimental sites

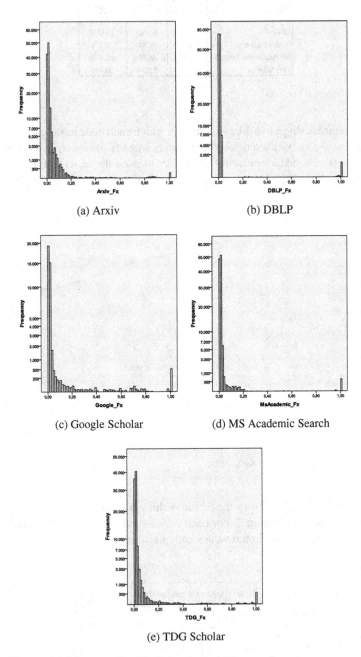

(a) Arxiv   (b) DBLP

(c) Google Scholar   (d) MS Academic Search

(e) TDG Scholar

Fig. 4 Histograms of feature values F_X for each site in the experiment

Just by looking at these statistics, we can see that the distributions are right-skewed, and that the vast majority of data are concentrated near the left extreme of the distribution (around 0).

For each site, we obtained the histograms in Figure 4, in which the Y-axis is expressed in power scale. We corroborate that we have two types of tokens, depending on their F_X value: on one hand we have a large share of data around 0, and on the other hand we have a small but significant share of data in the other extreme of the distribution, around 1. The first kind of tokens are those that appear rarely in other hubs different than that from which we obtained it. That is, their appearance depends on the query we made to obtain the hub, so they are similar to parameter values that are passed on to the server to create the hub. On the contrary, tokens around 1 are always present in every possible hub we obtain from the site, and they do not depend on the query.

As a consequence, we can use the information of the features to distinguish between those types of tokens, and create URL prototypes to classify Web pages using only their URL.

4   Conclusions and Future Work

In this paper, we present an experiment to proof that information contained in URL tokens is sufficient to classify those URLs, and to create classes of URLs.

We have developed features that exploit the information contained in the tokens. These features are probabilities estimators, based on token frequencies obtained from the experiment. The feature values histograms show that we can distinguish between two type of tokens, depending on their feature values, being some of them dependant on the query made to obtain the URL, whilst others are independent from it, and they always belong to URLs of the same type. As a conclusion, URL tokens indeed contain enough information to build URL prototypes, and therefore, to classify web pages, with the advantage of not having to download the page previously.

We have identified other proposals in the literature that aim at using the information contained in URLs to classify web pages. The advantages of our features in comparison with other proposals are 1) user intervention is kept to a minimum, which saves an important asset as is user time; 2) pages are classified for features that are outside them, which avoids having to download a page in order to classify it; 3) it is language independent, since it is based on the URL format regardless of the particular words or sequences of characters that make each token; 4) it does not require links to be surrounded by words useful for classification; and 5), we do not need to crawl extensively a site in order to build a classification model that works properly, instead, we perform a lightweight crawling that retrieves a small subset of pages. Because of the statistical nature of the proposal, we can be confident that the classifier is as accurate as it would be in case it had been built using the whole set of pages.

In the future, we plan to build a web page classifier using these features. We provide some insight about how to build such a classifier in [10]. Furthermore, we believe such a classifier can be used to improve web crawlers efficiency, which we expose in [9]. We must note that we must analyse how to apply our features to the so called friendly URLs, which do not fit our hypothesis.

References

1. Arasu, A., Garcia-Molina, H.: Extracting structured data from web pages. In: SIGMOD, pp. 337–348 (2003)
2. Bar-Yossef, Z., Rajagopalan, S.: Template detection via data mining and its applications. In: WWW, pp. 580–591 (2002)
3. Baykan, E., Henzinger, M.R., Marian, L., Weber, I.: Purely URL-based topic classification. In: WWW, pp. 1109–1110 (2009)
4. Baykan, E., Henzinger, M.R., Weber, I.: Web page language identification based on URLs. PVLDB 1(1), 176–187 (2008)
5. Blanco, L., Crescenzi, V., Merialdo, P.: Structure and semantics of Data-IntensiveWeb pages: An experimental study on their relationships. J. UCS 14(11), 1877–1892 (2008)
6. Blanco, L., Dalvi, N., Machanavajjhala, A.: Highly efficient algorithms for structural clustering of large websites. In: WWW, pp. 437–446. ACM, New York (2011)
7. Cohen, W.W.: Improving a page classifier with anchor extraction and link analysis. In: NIPS, pp. 1481–1488 (2002)
8. Fürnkranz, J.: Hyperlink ensembles: a case study in hypertext classification. Information Fusion 3(4), 299–312 (2002)
9. Hernández, I., Sleiman, H.A., Ruiz, D., Corchuelo, R.: A Conceptual Framework for Efficient Web Crawling in Virtual Integration Contexts. In: Gong, Z., Luo, X., Chen, J., Lei, J., Wang, F.L. (eds.) WISM 2011, Part II. LNCS, vol. 6988, pp. 282–291. Springer, Heidelberg (2011)
10. Hernández, I., Rivero, C.R., Ruiz, D., Corchuelo, R.: A Tool for Link-Based Web Page Classification. In: Lozano, J.A., Gámez, J.A., Moreno, J.A. (eds.) CAEPIA 2011. LNCS, vol. 7023, pp. 443–452. Springer, Heidelberg (2011)
11. Hotho, A., Maedche, A., Staab, S.: Ontology-based text document clustering. KI 16(4), 48–54 (2002)
12. Kan, M.-Y., Thi, H.O.N.: Fast webpage classification using URL features. In: CIKM, pp. 325–326 (2005)
13. Pierre, J.M.: On the automated classification of web sites. CoRR, cs.IR/0102002 (2001)
14. Selamat, A., Omatu, S.: Web page feature selection and classification using neural networks. Inf. Sci. 158, 69–88 (2004)
15. Vidal, M.L.A., da Silva, A.S., de Moura, E.S., Cavalcanti, J.M.B.: Structure-based crawling in the hidden web. J. UCS 14(11), 1857–1876 (2008)
16. Zhu, M., Hu, W., Wu, O., Li, X., Zhang, X.: User oriented link function classification. In: WWW, pp. 1191–1192 (2008)

On Relational Learning for Information Extraction

Patricia Jiménez, José Luis Arjona, and J.L. Álvarez

Abstract. The extraction and integration of data from multiples sources are required in current companies which manage their business process by heterogeneous collaborating applications. However, integrating web applications is an arduous task because they are intended for human consumption and they do not provide APIs to access to their data automatically. Web Information extractors are used for this purpose but, they mostly provide ad-hoc highly domain dependent solutions. In this paper we aim at devising Information Extractors with a FOIL based core algorithm. It is a widely used first order rule learning algorithm since their rules are substantially more expressive and allow to learn complex concepts that cannot be represented in the attribute-value format. Furthermore, we focus on integrating other scoring functions to check if we can improve the rule search guide speeding up the learning process in order to make FOIL tractable in real-world domains such as Web sources.

1 Introduction

The World Wide Web has become one of the largest repository of knowledge and the immediate standard to publish information. However, this information is offered via Hypertext Markup Language (HTML), what makes its perception easier for humans but not appropriate for automatic processing, since web sources do not usually

Patricia Jiménez
University of Sevilla, Avda. Reina Mercedes, 41012, Sevilla
e-mail: `patriciajimenez@us.es`

J.L. Arjona
University of Huelva (La Rábida), Palos de la Frontera 21071, Huelva
e-mail: `arjona@dti.uhu.es`

J.L. Álvarez
University of Huelva, La Rábida, Palos de la Frontera 21071, Huelva
e-mail: `alvarez@dti.uhu.es`

J.M.C. Rodríguez et al. (Eds.): Trends in PAAMS, AISC 157, pp. 117–124.
springerlink.com   © Springer-Verlag Berlin Heidelberg 2012

provide an Application Programmatic Interface (API) to interact with its interface automatically.

Web-Wrappers are the software component used for this purpose. They are intended to emulate the behaviour of a person who is interacting with a web user interface. It consists of an Enquirer, which maps user queries onto search forms, a Navigator, which executes filled search forms and reaches data web pages, an Information Extractor, which extract the information of interest from data web pages and return it stored as structured data for further processing. Finally, the Verifier attempts to find erroneous result sets. Our focus is on providing engineering support to devise Information Extractors.

Unfortunately, most Information Extractors today are inferred in a very ad-hoc way, in the sense of they can effectively extract information from a specific web site and achieve very good performance, but they may not be applied to other web sites with the same success. The ability to scale with the number and variety of information sources becomes the central challenge to information extraction (IE).

We wish to tackle this problem from inductive logic programming perspective. We aim at devising Information Extractors automatically with a FOIL [12, 13] based core algorithm. It is a widely used first order rule learning algorithm since their learnt rules are substantially more expressive, and allow the system to learn relational and recursive concepts that cannot be represented in the attribute-value format assumed by most machine learning algorithms. The application of relational learning can be decisive in domains that exhibit substantial variability such as Web pages.

Modified versions of FOIL are the basis for most adaptive IE systems that use relational learning techniques. For example, SRV system of Freitag [3] is one of the most successful ILP and top-down based learning system used for IE, which strongly follows the idea of the standard FOIL algorithm. The system is capable of learning extraction rules explaining single slots from natural and HTML documents. Moreover, Freitag extended SRV's feature predicates to make SRV able to exploit HTML structure by adding HTML-specific features. Although SRV almost always performs better than other learners, it does not solve all any new fields outright.

In order to deal with complex real-world domains for IE where the search space is not tractable, we work on devising an improved version of FOIL by applying some optimisations and heuristics. In this paper we present one of the optimisations consisting of replacing Information-based scoring function with other scoring functions coming from statistics, machine learning and data mining literature. We wish to find out through 16 ilp and classification tasks whether there is some scoring function that guides the search of the rules in a more efficient way, speeding up the learning process.

The paper is organised as follows: section 2, introduces an overview of FOIL algorithm. In section 3, we present some previous works on improving FOIL. Next, a common notation and a set of scoring functions are proposed. Then, we explain the experiments performed and we discuss the significance of our results. Finally, our future work is addressed in section 6.

2 FOIL Overview

Training data in FOIL comprises a target predicate, which is defined by a collection of positives and negatives examples according to whether they satisfy the target predicate or not, and a set of support predicates, which are defined extensionally, by a set of ground tuples.The goal is to learn a set of rules that explain the target predicate in terms of itself and the support predicates. The set of first order rules are represented as function-free Horn clauses and can optionally contain negated body literals.

It uses separate-and-conquer method rather than divide-and conquer, focusing on creating a single rule at a time and removing the positive examples covered by each learnt rule. Then, it is invoked again to learn a second rule based on the remaining training examples. It is called a sequential covering algorithm because it sequentially learns a set of rules that together cover the full set of positive examples. Additionally, FOIL employs a mechanism to speed up this process, pruning vast parts of the literal space when they show to be no better literals than the ones found so far.

In order to learn each rule, it follows a top-down approach, starting with a rule with an empty list of antecedents, and guided by a greedy search, the body of the rule is extended iteratively by adding the best new literals chosen according to a scoring function. This information-based scoring function, is designed to ensure that the learner will choose literals that include many positive examples and exclude many negative ones, while maintaining good overall coverage. Construction of a single rule stops if it matches only positive examples or reaches a predefined minimum accuracy. Furthermore, FOIL include MDL criterion [14] that stops the growth of the rule if the encoding length of the rule exceeds the number of bits needed for explicitly encoding the positive examples it covers. Thus, the induction of overly long and specific rules is prevented, especially in noisy-domains.

3 Related Work

Many authors have already tried to improve the performance of FOIL in several ways. Some earlier proposals are:

mFOIL [6] employs techniques from attribute-value learning to improve its noise handling capacities. It also offers two alternatives to the information-based scoring function, laplace-estimate and m-estimate. Moreover, FOIL's encoding length is replaced with criteria relying on statistical significance testing. Finally, it conducts beam-search to overcome, at least partially, some of the disadvantages of FOIL's greedy hill-climbing search. mFOIL is able to process intensionally defined background predicates and allows the user to define additional constraints to gain efficiency.

FOIDL [10] is also able to process intensionally defined background knowledge and negative examples are not needed. It assumes output completeness, i.e., the tuples in the relation show all valid outputs. Finally, it supports the induction of decision lists. That is an ordered set of rules each ending with a cut. When answering a query, the decision list returns the answer of the first rule in the ordered set which succeeds in answering the query. Rules are in reverse order, being the most general rules, those that cover many positive examples, at the end of the decision list.

FOCL [11] develops a hybrid method that combines inductive learning and explanation-based components. The latter allows advantageously to accept as input a partial, possible incorrect rule as an approximation of the target predicate. The candidate rules are evaluated using FOIL's information-based scoring function.As mFOIL, it also relies on user-defined constraint to restrict literal search space.

FOSSIL [2], uses a statistical correlation-based scoring function. It can be used to deal with noise by cutting off all literals that have a scoring function value below a certain threshold. They demonstrated that this threshold was independent of the number of training examples and of the amount of noise in the data. Moreover, they provide a new stopping criterion independent of the number of training examples and dependent on this statistical correlation scoring function.

The system nFOIL [8] integrates the nave Bayes learning scheme with FOIL. Two main changes on FOIL are required: first, examples that are already covered have still to be considered when learning additional rules; second, scoring functions is based on class conditional likelihood rather than information-based. nFOIL was shown to perform better than FOIL and to be competitive with more sophisticated approaches.

FZFOIL [4] uses interest-based measures to compute the score of a literal to overcome some lacks of Information scoring function and increase accuracy of the learnt rules. FZFOIL also manage fuzzy knowledge background predicates, where tuples are associated with these predicates with a certain degree of agreement. Induced rules are represented in both, ordinary and fuzzy logic format, and might generate incomplete and/or inconsistent rules. It process intensionally defined background predicates as well.

4 Our Contribution

We have implemented in Java an algorithm fairly similar to the last version of FOIL (FOILv6.4) which, unlike FOIL, it also supports defining background predicates intensionally, in the well-known Prolog-rules representation. It is possible through the JPL library that provides an interface between Java and Swi-Prolog. Some improvements need still to be incorporated in order to supply a more closer version of FOILv6.4.

This is a first attempt to check the behaviour of different scoring functions mainly taken from [15] and [7]. According to [4] we think that replacing information-based scoring function could improve the efficiency of the learning process at guiding the search, while retaining expressiveness.

The proposed scoring functions have been adapted according to the notation of the well-known confusion matrix, as appear in table 1. In any confusion matrix, tp (true positives) denotes the number of positive examples and fp (false positives) denotes the number of negative examples satisfied by a candidate literal. Similarly, fn (false negatives) and tn (true negatives) denote the number of positive and negative examples respectively, excluded by this literal. Finally, N is the total number of examples in the current training set.

Table 1 List of Scoring functions

Scoring Function	Formula	Scoring Function	Formula
Information (I)	$-log\frac{tp}{tp+fp}$	Laplace Accuracy (Lap)	$\frac{tp+1}{tp+fp+2}$
Leverage (Lev)	$\frac{tp\cdot tn - fn\cdot fp}{N^2}$	ϕ-coefficient (ϕ)	$\frac{tp\cdot tn - fn\cdot fp}{\sqrt{(tp+fn)\cdot(tp+fp)\cdot(fp+tn)\cdot(fn+tn)}}$
Confidence (Conf)	$\frac{tp}{tp+fp}$	Satisfaction (Sat)	$\frac{tp\cdot tn - fp\cdot fn}{(tp+fp)\cdot(tn+fp)}$
F-measure (F_1)	$2\cdot\frac{tp}{2\cdot tp+fn+fp}$	kappa (κ)	$\frac{2\cdot(tp\cdot tn - fp\cdot fn)}{N^2-(tp+fn)\cdot(tp+fp)-(fp+tn)\cdot(tn+fn)}$
Odds-ratio (OR)	$\frac{tp\cdot tn}{fp\cdot fn}$	Yule's Q (Q)	$\frac{tp\cdot tn - fp\cdot fn}{tp\cdot tn + fp\cdot fn}$
Lift (L)	$\frac{N\cdot tp}{(tp+fp)\cdot(tp+fn)}$	Jaccard(ζ)	$\frac{tp}{tp+fp+fn}$
Collective Strength (CS)	$\frac{tp+tn}{(tp+fp)\cdot(tp+fn)+(fn+tn)\cdot(fp+tn)} \times \frac{N^2-(tp+fp)\cdot(tp+fn)-(fn+tn)\cdot(fp+tn)}{N-tp-tn}$		

5 Experiments

To carry out our analysis, we have performed 16 experiments taken from ILP, machine learning and classification problems literature. Each learning task involved a limited amount of background information, just that required for the task at hand, and training examples noise-free.

Amongst the trials carried out, we can highlight kindship from Hinton [5], arch task, introduced by Winston [16], Michalski East-West trains problem [9], play tennis and contact lenses classification problems taken from [17] and several Ivan Bratko's tasks on a universe of three-length lists taken from well-known text Prolog Programming for Artificial Intelligence [1].

Table 2 Results

Tests		I	Lap	Lev	ϕ	Conf	Sat	F_1	κ	OR	Q	Lift	J	CS
member	R	1.00	1.00	1.00	1.00	1.00	1.00	1.00	1.00	1.00	1.00	1.00	1.00	1.00
75+/120	T	3.20	3.74	3.20	3.85	3.59	3.23	3.62	3.45	3.48	20.84	3.45	3.46	3.60
	C	8.90	8.90	8.09	8.90	8.90	8.90	8.90	8.90	8.90	8.90	8.90	8.90	8.90
del	R	1.00	1.00	1.00	1.00	1.00	1.00	1.00	1.00	1.00	time	1.00	1.00	0.56
81+/4800	T	137.2	124.1	184.4	220.1	125.8	117.9	135.8	130.6	172.6	-	119.8	132.4	185.4
	C	14.57	14.57	24.43	14.57	14.57	14.57	14.57	14.57	14.57	-	14.57	14.57	19.82
last	R	1.00	1.00	1.00	1.00	1.00	1.00	1.00	1.00	1.00	1.020	1.00	1.00	1.00
39+/120	T	5.64	5.55	5.38	5.71	5.67	5.48	5.40	5.37	16.58	11.06	17.63	5.35	3.76
	C	9.17	9.17	9.17	9.17	9.17	9.17	9.17	9.17	10.49	9.75	10.49	9.17	8.17
insert	R	1.00	1.00	1.00	1.00	1.00	1.00	1.00	1.00	1.00	time	1.00	1.00	1.00
81+/4800	T	130.7	118.3	170.5	195.6	117.9	117.3	129.2	128.6	166.4	-	120.7	130.3	84.13
	C	14.19	14.19	23.96	14.19	14.19	14.19	14.19	14.19	14.19	-	14.19	14.19	14.40
sublist	R	1.00	1.00	1.00	1.00	1.00	1.00	1.00	1.00	1.00	1.00	1.00	1.00	1.00
202+/1600	T	30.64	24.84	30.28	59.03	29.52	29.83	63.48	24.32	48.34	95.16	23.51	24.82	83.98
	C	12.71	12.71	12.71	12.71	12.71	12.71	14.75	12.71	12.71	14.75	12.71	12.71	26.33
even	R	1.00	1.00	1.00	1.00	1.00	1.00	1.00	1.00	1.00	1.00	1.00	1.00	1.00
10+/40	T	0.83	0.76	0.72	0.78	0.73	0.81	0.83	0.80	1.72	1.48	1.46	0.76	0.73
	C	5.00	5.00	5.00	5.00	5.00	5.00	5.00	5.00	5.00	17.36	17.36	5.00	5.00
permutation	R	1.00	1.00	1.00	1.00	1.00	1.00	1.00	1.00	1.00	1.00	1.00	1.00	1.00
52+/256	T	40.58	36.96	7.86	23.18	34.06	254.5	5.87	7.85	22.04	193.6	250.7	5.82	9.19
	C	17.97	17.97	22.44	29.68	17.97	49.69	25.72	22.44	29.53	26.38	68.28	25.72	22.72
playtennis	R	1.00	1.00	0.57	0.79	1.00	0.43	0.50	0.57	0.43	0.43	0.14	0.50	0.71
14+/72	T	7.22	7.18	13.30	13.84	8.61	3.71	7.12	5.52	5.65	4.35	1.84	6.80	11.82
	C	172.0	170.4	107.2	137.8	170.2	50.19	70.00	66.17	52.28	51.82	18.40	70.00	118.1
lenses	R	1.00	1.00	0.50	1.00	1.00	1.00	0.58	0.50	0.50	0.50	0.58	0.58	1.00
24+/72	T	3.68	3.74	1.61	6.72	4.02	4.88	3.04	1.56	1.58	1.37	2.30	2.98	11.10
	C	102.7	102.7	8.76	111.5	102.7	101.4	24.01	8.76	8.76	8.76	24.79	24.01	112.3
plus	R	1.00	1.00	-	1.00	1.00	1.00	-	1.00	1.00	-	1.00	-	1.00
6+/27	T	1.95	2.01	-	6.11	2.14	2.00	-	2.11	4.91	-	7.02	-	8.02
	C	18.06	18.06	-	17.06	18.06	18.06	-	18.06	19.47	-	22.76	-	15.66
mult	R	1.00	1.00	1.00	1.00	1.00	1.00	1.00	1.00	1.00	1.00	1.00	1.00	1.00
48+/1056	T	43.38	34.07	77.63	111.6	35.54	33.04	41.39	39.37	65.24	221.8	36.66	40.05	24.04
	C	29.10	29.10	29.58	29.10	29.10	29.10	29.10	29.10	29.10	29.77	29.10	29.10	16.49
animals	R	1.00	1.00	1.00	1.00	1.00	1.00	1.00	1.00	0.29	0.29	1.00	1.00	1.00
17+/68	T	4.02	3.66	4.35	7.41	4.00	4.29	5.91	4.41	2.39	2.23	4.26	5.88	3.09
	C	18.81	18.81	18.81	21.81	18.81	18.81	18.81	18.81	7.00	8.17	18.81	18.81	21.42
network	R	1.00	1.00	1.00	1.00	1.00	1.00	1.00	1.00	1.00	-	1.00	1.00	1.00
19+/81	T	0.86	1.39	0.98	1.01	1.59	1.17	5.49	1.65	0.83	-	1.39	5.71	0.81
	C	12.34	12.34	12.34	13.92	12.34	12.34	42.05	12.34	12.34	-	12.34	42.05	12.34
kindship	R	1.00	1.00	1.00	1.00	1.00	1.00	1.00	1.00	1.00	1.00	1.00	1.00	1.00
12+/576	T	6.21	5.85	6.54	9.61	5.76	5.82	6.13	6.55	8.16	8.22	6.04	6.30	6.36
	C	9.49	9.49	9.49	9.49	9.49	9.49	9.49	9.49	9.49	9.49	9.49	9.49	26.68
trains	R	1.00	1.00	1.00	1.00	1.00	1.00	1.00	1.00	1.00	1.00	1.00	1.00	1.00
5+/10	T	0.72	0.64	0.72	0.76	0.72	0.72	0.73	0.75	0.73	0.83	0.80	0.76	0.72
	C	8.82	8.82	8.82	8.82	8.82	8.82	8.82	8.82	8.82	8.92	8.92	8.82	8.82
arch	R	1.00	1.00	1.00	1.00	1.00	1.00	1.00	1.00	1.00	-	1.00	1.00	1.00
2+/1728	T	6.60	5.69	6.83	8.44	6.29	6.18	5.83	6.07	9.38	-	9.08	6.19	8.33
	C	19.97	19.97	19.97	19.97	19.97	19.97	19.97	19.97	19.97	-	19.97	19.97	19.97

The adopted measures to help us to compare the results are[1]:

1. **Recall (R)**. It determines if a set of rules is complete, i.e., if it satisfies all positive examples belonging to the target predicate.
2. **Complexity (C)** of the induced set of rules. It is computed in terms of bits from Minimum Description Length Principle [14].
3. **Time (T)** employed for rules learning process measured in seconds and restricted to 1000s.

The results in table 2 exhibit that in the 81,25% of cases there is some scoring function that performs better than the information-based one, in terms of time and complexity, maintaining the full recall. Information-based, Laplace, Confidence and Satisfaction scoring functions seem to be the most promising scoring functions. Nevertheless, they do not always provide the best results. In a few cases, Leverage and Collective Strength obtained best times inducing the same or similar rules that the ones induced applying information-based scoring function. Inducing rules in shorter time may be due to both, a more effective alpha-beta pruning and search guide, which depend completely on the scoring function selected.

Note that Yule's Q scoring function is the worst employed. It is not able to induce a set of rules or it wastes a lot of time to induce them. Neither OR scoring function provided results to take into account, only in one case behaved not much more speedy than applying information-based scoring function. We also should know that Collective Strength and ϕ-coefficient do not have implemented the alpha beta pruning yet, because it requires significant changes in the implementation. But even competing at a disadvantage, the former gets to be the most promising one in a 18,75% of cases. Unfortunately, the latter produced no significant results. It exceeded the time achieved by information based-scoring function in the 93,75% of cases. Finally, Lift and Jaccard scoring functions also got better results than those obtained with information-based scoring function in a 31,75% and 43,75% of cases respectively. However, they never were the most promising ones in any task.

Clearly, the results stated that most of these scoring functions are not recommendable for classification tasks which are the playTennis and lenses tests. Only Information Gain, Laplace and Confidence reached a full recall. In other tasks as plus, insert or del, there were some functions that could not induced the set of rules by time constraints or simply because they didn't find the rules.

6   Conclusions

We have implemented a customised version of the well-known FOIL algorithm and we have proposed to integrate different scoring functions taken from the literature, in order to guide the search for a rule efficiently.

Thirteen scoring functions were applied over 16 tests from ILP and classification domain in order to compare them. Although many of these scoring functions performed reasonable well, the experiment highlights the strong dependency between the task and the scoring function applied, since they got good results in some tasks

---

[1] Note that we do not allow to cover negative examples so, rules are 100%accurate.

124   P. Jiménez, J.L. Arjona, and J.L. Álvarez

and not so well in others. For this reason, we think of applying some algorithm to
rank them as in [15] and decide the scoring function that best fits for a specific task.

As future work, we plan include new adapted scoring functions for extending the
possibilities. Next steps are addressed to provide new optimisations and heuristics
that make our approach tractable in real-world applications related to IE.

Acknowledgements. This research is supported by the European Commission (FEDER), the
Spanish and the Andalusian R&D&I programmes (grants TIN2007-64119, P07-TIC-2602,
P08-TIC-4100, and TIN2008-04718-E).

References

1. Bratko, I.: Prolog Programming for Artificial Intelligence. In: McGettrick, A.D., Van
   Leeuwen, J. (eds.). Addison-Wesley (1986)
2. Fürnkranz, J.: FOSSIL: A Robust Relational Learner. In: Proc. of the Eur. Conf. on Mach.
   Learn. (1994), doi:10.1007/3-540-57868-4_54
3. Freitag, D.: Information Extraction from HTML: Application of a General Machine
   Learning Approach. In: Proc. Fifteenth Natl. Conf. on Artif. Intell., pp. 517–523 (1998)
4. Gomez, A.J., Fernandez, G.: Induccion de definiciones logicas a partir de relaciones:
   mejoras en los heuristicos del sistema FOIL. In: Congr. Nac. Program. Declar., pp. 292–
   302 (1992)
5. Hinton, G.E.: Learning distributed representations of concepts. In: Proc. of the Eighth
   Annu. Conf. of the Cogn. Sci. Soc., pp. 1–12 (1986)
6. Lavrac, N., Dzeroski, S.: Inductive Logic Programming: Techniques and Applications.
   In: Lavrac, N., Dzeroski, S. (eds.) Inductive Logic Programming, pp. 173–179. Hellis
   Horwood, New York (1994)
7. Lavrac, N., Flach, P.A., Zupan, B.: Rule Evaluation Measures: A Unifying View. In:
   Proc. of the 9th Int. Workshop on Inductive Log. Program. (1999), doi:10.1007/3-540-
   48751-4_17
8. Landwehr, N., Kersting, K., De Raedt, L.: nFOIL: Integrating Naïve Bayes and FOIL.
   In: The 20th Natl. Conf. on Artif. Intell., pp. 795–800 (2005)
9. Michalski, R.S.: Pattern recognition as rule-guided inductive inference. IEEE Trans. on
   Pattern Analysis and Mach. Intell. 2, 349–361 (1980)
10. Muggleton, S.: Inverse Entailment and Progol. New Gener. Comput. J. (1995),
    doi:10.1007/BF03037227
11. Pazzani, M.J., Kibler, D.F.: The Utility of Knowledge in Inductive Learning. Mach.
    Learn. 9, 57–94 (1992)
12. Quinlan, J.R., Cameron-Jones, R.M.: FOIL: A Midterm Report. In: Proc. of the Eur.
    Conf. on Mach. Learn. (1993), doi:10.1007/3-540-56602-3_124
13. Quinlan, J.R., Cameron-Jones, R.M.: Induction of Logic Programs: FOIL and Related
    Systems. New Gener. Comput. J. 13, 287–312 (1995)
14. Rissanen, J.: Universal coding, information, prediction, and estimation. IEEE Trans. Inf.
    Theory 30, 629–636 (1984)
15. Tan, P., Kumar, V., Srivastava, J.: Selecting the right objective measure for association
    analysis. Inf. Syst. (2004), doi:10.1016/S0306-4379(03)00072-3
16. Winston, P.H.: Learning Structural Descriptions from Examples. In: Winston, P.H. (ed.)
    The Psychology of Computer Vision, pp. 157–209. McGraw-Hill, New York (1975)
17. Witten, I.H., Frank, E., Hall, M.A.: Data Mining: Practical machine learning tools and
    techniques with Java implementations, pp. 9–13. Morgan Kauffman (2000)

Automatic Optimization of Web Navigation Sequences

José Losada, Juan Raposo, Alberto Pan, and Javier López

Abstract. Web automation applications are widely used for different purposes such as B2B integration, automated testing of web applications or technology and business watch. In this work-in-progress paper we outline a set of techniques which constitute the basis to build a web navigation component able to analyze a web navigation sequence and automatically optimize it, detecting which parts of the loaded pages are needed, and which ones can be discarded in the following executions of the sequence. Our techniques build on the Document Object Model and the first tests executed with real web sources have found them to be very effective.

1 Introduction

Web automation applications are widely used for different purposes such as B2B integration, web mashups, automated testing of web applications, Internet meta-search or technology and business watch. One crucial part in web automation applications is how to easily generate and reproduce navigation sequences. We can identify two distinct stages in this process:

- In the *generation* stage the user specifies the navigation sequence to reproduce. The most common approach, cf. [1, 6, 7, 9], is using the 'recorder' metaphor.
- In the *execution* phase the sequence generated in the previous stage is provided as input to an automatic navigation component which is able to reproduce it.

The automatic navigation component used in the execution phase can be developed by using the APIs of popular browsers, cf. [4, 7, 8, 10, 11], or simplified custom browsers specially built for the task, cf. [1, 3, 5, 9].

José Losada · Juan Raposo · Alberto Pan · Javier López
Information and Communications Technology Department, University of A Coruña
Facultad de Informática, Campus de Elviña, s/n, 15071, A Coruña (Spain)
e-mail: {jlosada,jrs,apan,jmato}@udc.es

J.M.C. Rodríguez et al. (Eds.): Trends in PAAMS, AISC 157, pp. 125–132.
springerlink.com   © Springer-Verlag Berlin Heidelberg 2012

The approach of using the APIs of commercial web browsers have some advantages: no effort is required to develop a navigation component, and the accessed web pages behave the same as when they are accessed by a regular user navigating with the commercial web browser. But, on the contrary, the performance of the component is limited by the commercial browser performance and the functionalities that the browser API provides. The main purpose of commercial web browsers is to be used by human users, and they consume a significant amount of resources, both memory and CPU. So, this approach is not the most appropriate to execute intensive or real time web automation tasks, which need to execute a significant number of navigation sequences in the less possible time.

The approach of creating a custom browser, supporting technologies such as scripting code and AJAX requests, is effort-intensive and can be vulnerable to implementation differences that can make a web page behave differently when accessed with the custom browser. Nevertheless, the main advantage is the performance: custom browsers can consume fewer resources (memory and CPU), and they are able to execute navigation sequences faster than commercial browsers.

Current systems which use the approach of creating custom browsers to execute navigation sequences, like [3] or [5], can avoid some steps executed by commercial web browsers (e.g. the page rendering phase), but they replicate its functioning when loading and building the internal representation of the web pages. The pages are always completely loaded (e.g. all the scripts contained in the page are executed).

In this work-in-progress paper we present the basis for a web navigation component able to analyze a web navigation sequence and automatically optimize it, detecting which parts of the loaded pages can be discarded:

- In the *optimization phase* the sequence is executed once, and in the meantime the execution component automatically calculates which nodes of the HTML DOM [2] tree of the loaded pages are needed to execute the sequence and which ones can be discarded. Then, it stores some information to be able to detect those elements in subsequent sequence executions.
- In the *execution* phase the execution component executes the sequence using the optimization information. When each page is loaded, a reduced HTML DOM tree is built, containing only the relevant nodes needed to execute the sequence.

This way, smaller HTML DOM trees are built when each page is loaded resulting in less memory usage. Besides, the script code, including AJAX requests, contained in elements not loaded in the simplified tree are not executed, and the external resources (e.g. JavaScript or CSS files) referenced from elements not loaded in the tree are not retrieved, therefore optimizing CPU time and network usage.

The step of searching which elements must be loaded in the simplified HTML DOM tree is the unique latency that the navigation component adds at execution time. As we will demonstrate in the experimental evaluation, this latency is insignificant compared to the time savings derived from building the simplified tree.

2 Models

In this section we briefly describe the model we use to characterize the component used to automatically optimize web navigation sequences.

The main model we rely on is the Document Object Model (DOM) [2]. This model describes how browsers internally represent the HTML web page currently loaded in the browser and how they respond to user-performed actions on it. An HTML page is modelled as a tree, where each HTML element is represented by an appropriate type of node. An important type of nodes are the script nodes, used to place and execute a script code within the document (typically written in a script language such as JavaScript). The script nodes can contain the script code directly or can reference an external file containing it. Those scripts are processed when the page is loaded and they can contain element declarations (e.g. a function or a variable) that are used from other script nodes or event listeners.

Each node in the tree can receive events produced (directly or indirectly) by the user actions. Event types exist for actions such as clicking on an element (*click*), moving the mouse cursor over it (*mouseover*), or to indicate that a new page has just been loaded (*load*), to name but a few. Each node can register a set of event listeners for different types of events. Each event is dispatched following a path from the root of the tree to the target node, and it can be handled locally at the target node or at any target's ancestors in the tree (this is called "bubbling"). An event listener executes arbitrary code, which normally calls a function declared in script nodes.

In the context of the Document Object Model, we say that there exists a dependency between two nodes N_1 and N_2 when the node N_1 is necessary for the correct execution of the node N_2. We say that the node N_1 is a dependency of the node N_2. The following rules define the dependencies involving nodes which execute script code (script nodes or nodes containing event listeners):

1. If the script code of a node S_1 uses an element (e.g. a function or a variable) declared in a script node S_2, then S_2 is a dependency of S_1. To be able to execute the script code of the node S_1 the node S_2 must be loaded.
2. If the script code of a node S uses a node N (e.g. using the JavaScript function *document.getElementById*), then N is a dependency of S. To be able to execute the script code of the node S, the node N must be loaded.
3. If the script code of a node S makes a modification in a node N (e.g. it modifies the *action* attribute of a form node), then S is a dependency of N. If the node N is going to be used, then the script node S needs to be loaded to perform the modification in the node N.

Note that, node dependencies are transitive. For example, if an event listener of a node N_1 invokes a function f which is defined in a node N_2, and the implementation of f uses the node N_3, then both N_2 and N_3 are dependencies of N_1.

3 Description of the Solution

3.1 Optimization Phase

The optimization phase involves one execution of the navigation sequence where the navigation component automatically calculates which nodes of the DOM trees of the loaded pages are needed to execute the sequence (*relevant* nodes), and which ones can be discarded (*irrelevant* nodes). Then, it stores some information to be able to identify these nodes in the following executions of the sequence.

First, we will explain the techniques designed to calculate the set of *relevant* nodes and the set of *irrelevant* nodes. While executing a navigation sequence we can differentiate two steps for each page loaded:

1. The *page loading* step involves loading the page, generating the DOM tree, downloading external elements (e.g. style sheets, script files) and executing the script nodes defined in the page. Finally, some predefined events are automatically fired when the new page is completely loaded (e.g. the *load* event is fired over the body node), and some event listeners can be executed as response.
2. The *page interaction* step involves executing the pertinent actions and firing the necessary events, to execute the navigation sequence commands which emulate the user interaction with the page (e.g. clicking on elements, firing mouse movement events, etc.), until a navigation to a new page is started.

In both steps, there may be multiple interactions among nodes in the page, which must be taken in consideration to determine which nodes of the DOM tree are required for the correct execution of the navigation sequence.

During the page loading step, the navigation component can use the rules explained in section 2 to build a node dependency graph, containing the node dependencies for all the script nodes that are executed, and for all the nodes which contain event listeners that are executed as response to the events fired. In a similar way, during the page interaction step, for each event which is fired, a dependency graph is calculated for all the nodes which execute event listeners in response to the event. Finally, all these node dependency graphs are merged into a unique global dependency graph.

Then, the set of relevant nodes can be built using the following rules:

1. The target nodes of each of the actions to be executed or the events to be fired during the page interaction step are relevant.
2. If a node is relevant, all its ancestors are relevant. This is needed because of the "bubbling" stage in the DOM event execution model (see section 2).
3. By definition, if a node is relevant, all its dependencies are relevant.
4. If an input node is relevant, the form node containing it is relevant.
5. If a form node is relevant, all the input and select nodes contained in the form are relevant.
6. If a select node is relevant, all its child option nodes are relevant (this rule and the two previous ones are needed to be able to properly submit forms).
7. A small set of node types are always considered relevant (e.g. base nodes).

To calculate the set of irrelevant nodes, first, all the DOM tree nodes not contained in the set of relevant nodes are added to it. Then, all the irrelevant nodes which have an ancestor also contained in the set of irrelevant nodes are removed from the set. The resulting set contains only the root nodes of the sub-trees whose descendants are all irrelevant (we call them irrelevant sub-trees).

Now, we will briefly explain the techniques used to generate expressions to identify the root nodes of the irrelevant sub-trees at execution time. On one hand, the generated expressions should be resilient to small changes in the page because in real web sites there are usually small differences between the DOM tree of the same page loaded at different moments (e.g. different data records can be shown in dynamically generated sections). On the other hand, the process of testing if a node matches an expression should be very efficient, because, at the execution phase the browser should check if each node matches with any of that expressions before creating and adding it to the HTML DOM tree.

To uniquely identify a node in the DOM tree we use an XPath-like expression. XPath [12] expressions allow identifying a node in a DOM tree by considering information such as the node type, the text associated to the node, the value of its attributes and its ancestors. Our proposal starts from a very simple XPath-like expression using only the type, text and attributes associated to the target node and, if it does not uniquely identify the node, the expression is progressively augmented including information from some appropriate ancestors, until it does. We build the least restrictive expression that still uniquely identifies the target node. Besides, these type of expressions can be evaluated efficiently at the execution phase. The algorithm is not deeply described due to space constraints.

3.2 Execution Phase

The general functioning of the navigation component at this phase is the following one: before loading each page, it checks if it has optimization information associated to that page, that is, a set of expressions to identify the root nodes of the page irrelevant sub-trees. That information is used during the parsing stage to build a reduced version of the page HTML DOM tree, containing only the relevant fragments. If a node matches with any expression of the set, the node is not added to the tree and the entire page fragment below that node is completely discarded.

The process of checking if a node is the root of an irrelevant sub-tree should be very efficient because it is executed for all the elements present in the page to decide if they must be added to the HTML DOM tree or not. Due to the method used to create the XPath-like expressions that identify the root nodes of the irrelevant sub-trees, an efficient algorithm has been designed to check if a node matches with an expression. The algorithm is not described due to space constraints.

Another important issue to deal with during the execution phase, is the identification of the pages where the optimizations should be applied. In most cases, the order in which the pages are loaded when the navigation sequence is executed could be used to identify them, but further investigation is required to design a more robust method for identifying the pages. This task is currently in progress.

4 Evaluation

To evaluate the validity of our approach a custom browser was implemented. This browser emulates Microsoft Internet Explorer (MSIE) version 8 and was fully implemented in Java using open-source libraries including Apache Commons-Httpclient to handle HTTP requests, Neko HTML parser to build DOM structures, and Mozilla Rhino as JavaScript engine. The custom browser has proved very efficient. It works in most of the navigation sequences and it is faster than MSIE and other commercial browsers (like Firefox, Chrome, etc.) is most of the cases. To evaluate the techniques and algorithms proposed in this paper they have been implemented in the core of this custom browser.

This section explains the preliminary set of experiments that we have designed and executed, which can be divided in two different types. For the first type, we selected a set of popular websites of different domains. In each website we recorded a navigation sequence involving several pages. We ran a first execution of the navigation sequence to collect the optimization information. Then, we ran two more executions, the first one without using the optimization information, and the second one using it. Table 1 shows the metrics measured for each of this two executions in a representative subset of the selected websites. (each cell shows the result of the normal execution followed by the result of the optimized one). Note that one execution is enough to calculate these metrics because they will have the same values in all the executions while the website pages remain without changes.

The results of the first type of experiments show that in almost every source more than the 50% of the nodes are identified as irrelevant. In half the sources, the nodes identified as irrelevant are more than the 75% (up to 96,5%) of the total nodes. Those irrelevant nodes include scripts, style sheets and frames. Discarding those nodes, the browser also avoids unnecessary downloads and the execution of unnecessary scripts, so the memory and CPU usage required to execute the navigation sequence is highly minimized when the optimization information is used.

The second type of experiments consists of a benchmark using 5 instances of our custom browser running in parallel, executing the same navigation sequence during a fixed amount of time (10 minutes). To avoid the latency of the network, some of the websites used in the first type of experiments were replicated in a local web server, simulating the original website. Table 2 shows the number of executions completed, using and without using the optimization information.

The results of the second type of experiments show that the executions using the optimization information are, in average, 41,7% (it varies from 21,7% to 70%) faster than normal executions. Note that these experiments use the sources replicated locally, so they do not include the time savings derived from downloading fewer resource files (CSS, JavaScript, etc.) from remote servers.

Table 1 Metrics comparing normal and optimized executions in some websites

Website	HTML DOM Nodes created	Scripts Executed	Frames and Windows	HTML pages Downloaded	External objects Downloaded	AJAX Requests
Reuters	3264 / 1463	303 / 168	6 / 3	9 / 7	176 / 103	4 / 4
Pixmania	3808 / 1527	156 / 96	2 / 1	5 / 5	75 / 46	1 / 0
Optize	2699 / 734	102 / 53	1 / 1	3 / 3	36 / 25	0 / 0
Wikipedia	4742 / 1168	69 / 58	5 / 1	5 / 5	47 / 43	4 / 4
Amazon	8319 / 5046	295 / 201	28 / 13	30 / 15	63 / 37	9 / 7
Ebay	5474 / 2603	119 / 111	10 / 8	14 / 13	33 / 31	0 / 0
Vueling	37865 / 7237	3504 / 894	178 / 31	78 / 28	1115 / 340	123 / 3
Bloomberg	6585 / 1160	351 / 212	8 / 3	12 / 7	162 / 103	2 / 0
Fnac	6993 / 1309	232 / 107	13 / 7	23 / 15	104 / 54	0 / 0
AppleStore	1914 / 67	40 / 17	1 / 1	3 / 3	12 / 11	0 / 0
NYTimes	6911 / 2016	279 / 223	8 / 5	14 / 11	192 / 155	4 / 4
Imdb	5033 / 1633	285 / 166	41 / 6	47 / 11	112 / 75	6 / 6
CNet	6117 / 1067	247 / 168	10 / 6	15 / 11	116 / 90	0 / 0
AbeBooks	3553 / 730	172 / 104	6 / 2	8 / 5	82 / 71	2 / 2
AllBooks4Less	2959 / 545	69 / 30	6 / 2	9 / 6	20 / 11	9 / 3

Table 2 Benchmarking normal versus optimized executions

☐	Pixmania	Optize	Wikipedia	Amazon	Ebay	Vueling
Normal	111	390	244	130	173	18
Optimized	195	635	323	277	218	60

5 Related Work

Currently, web automation applications are widely used for different purposes. The automatic navigation component used by these applications is developed by using the APIs of popular browsers or simplified custom browsers specially built for the task. WebVCR [1] and WebMacros [9] rely on simple HTTP clients that lack the ability to execute complex scripting code or to support AJAX requests. Wargo [7], Smart Bookmarks [4], Sahi [10], Selenium [11] and QEngine [8] use a commercial browser as execution engine. Therefore, the performance of the component is limited by the commercial browser performance, which consumes a significant amount of resources. HtmlUnit [3] and Kapow [5] use their own custom browser with support for many JavaScript and AJAX functionalities. They are more efficient than commercial web browsers, but they replicate its functioning when loading pages and building its internal representation, without allowing any type of extra optimizations.

6 Conclusions and Future Work

In this paper, we have presented a novel set of techniques and algorithms to optimize automatic web navigation sequences. Our approach is based on executing the navigation sequence once, to automatically collect information about the elements of the loaded pages that are irrelevant for that navigation sequence. Then, that information is used in the next executions of the sequence, to load only the required elements. According to a preliminary set of experiments they seem to be very effective, but further experimentation is required. We also plan to refine some of the algorithms like, for example, the method used to identify the pages that are loaded in order to apply them the correct optimizations.

Acknowledgments. This research was partially supported by the Spanish Ministry of Science and Innovation under project TIN2010-09988-E, and the European Commission under project FP7-SEC-2007-01 Proposal N° 218223.

References

[1]   Anupam, V., Freire, J., Kumar, B., Lieuwen, D.: Automating web navigation with the WebVCR. In: WWW 2000, pp. 503–517 (2000)

[2]   Document Object Model (DOM), http://www.w3.org/DOM/

[3]   HtmlUnit, http://htmlunit.sourceforge.net/

[4]   Hupp, D., Miller, R.C.: Smart Bookmarks: automatic retroactive macro recording on the web. In: ACM Symposium on User Interface Software and Technology (UIST), pp. 81–90 (2007)

[5]   Kapow, http://www.openkapow.com

[6]   Little, G., Lau, T., Cypher, A., Lin, J., Haber, E., Kandogan, E.: Koala: Capture, Share, Automate, Personalize Business Processes on the Web. In: SIGCHI Conference on Human Factors in Computing Systems, pp. 943–946 (2007)

[7]   Pan, A., Raposo, J., Álvarez, M., Hidalgo, J., Viña, A.: Semi automatic wrapper-generation for commercial web sources. In: IFIP WG8.1 Working Conference on Engineering Information Systems in the Internet Context, pp. 265–283 (2002)

[8]   QEngine,
       http://www.adventnet.com/products/qengine/index.html

[9]   Safonov, A., Konstan, J., Carlis, J.: Beyond Hard-to-Reach Pages: Interactive, Parametric Web Macros. In: 7th Conference on Human Factors & the Web (2001)

[10]   Sahi, http://sahi.co.in/w/

[11]   Selenium, http://seleniumhq.org/

[12]   XML Path Language (XPath), http://www.w3.org/TR/xpath

Metabolic Pathway Data and Application Integration[*]

Ismael Navas-Delgado, Maria Jesús García-Godoy, and José F. Aldana-Montes

Abstract. This paper shows three previous approaches for data integration, Linked Data access and Web Service annotation, and what problems have to be solved in the context of Life Sciences to integrate and use Metabolic Pathway data published as Linked Data.

1 Introduction

The amount of information on the World Wide Web has increased enormously over the last few years, but this information is still shown through HTML and other human-oriented contents. There has been a lot of effort to provide data sets from different areas using Semantic Web technologies, and Linked Data proposes a set of best practices for exposing, sharing and connecting data, information and knowledge by using RDF/S and URIs. This movement towards the publication and linking of data instead of contents has been continuously growing reaching around 31 billion RDF triples, which are interlinked by means of around 504 million RDF links. There have been some trials to improve the application of this available technology in biology areas. To address this, Bio2RDF started in 2008 [1] as a semantic Web application designed to solve the integration problem in the area of bioinformatics. Linked Life Data [2] is another platform that stores billions of RDF statements about biomedical knowledge.

On the other hand, Web services have emerged as a key technology in the development of distributed applications. Life sciences is a domain in which Web services have been widely adopted, and numerous approaches are making use of

Ismael Navas-Delgado · Maria Jesús García-Godoy · José F. Aldana-Montes
Universidad de Málaga, Bulevard Louis Pasteur nº35, 29071, Málaga, Spain
e-mail: {ismael,mjgarciag,jfam}@lcc.uma.es

[*] The Project Grants TIN2011-25840 (Spanish Ministry of Innovation, Science and Technology) and P11-TIC-7529 (Innovation, Science and Enterprise Ministry of the regional government of the Junta de Andalucía) have supported this work.

J.M.C. Rodríguez et al. (Eds.): Trends in PAAMS, AISC 157, pp. 133–140.
springerlink.com   © Springer-Verlag Berlin Heidelberg 2012

them to provide data retrieval and for analysis. Parallel to the increasing number of Web services available (in general but specifically in bioinformatics), the Semantic Web has also emerged. Thus, the extensive use of Web services for providing bioinformatics tools at the same time encountered a way of registering these tools to enable their use in complex tools and workflows. However, the emergence of Semantic Web technologies has not been widely taken up by the bioinformatics community for the annotation of these Web services. So, the sustained growth of available services in the Life Sciences has led to an explosion of bioinformatic Web Service registries.

This paper shows three approaches for taking advantage of these data and services to enable new scientific findings. Then, we focus on how these approaches can be combined to reach new solutions, and the main problems to be studied.

2 Related Work

The need for data integration started when the number of applications and data repositories began to grow rapidly. The first approaches appeared in the80s, and formed the basis for the research in this area. The evolution continued over mediator based systems, such as AMOS II [3], DISCO[4], TSIMMIS [5] and Garlic[6]. Then, agent technology was used in some systems like InfoSleuth[7] and MOMIS [8]. Finally, the new technologies appearing have been used in data integration: XML (MIX[9]), ontologies (OBSERVER [10]). Finally, more sophisticated and even commercial systems have appeared, such as FeDeRate [11], Virtuoso, SDS[1], and Semantic Web Middleware for Virtual Data Integration on the Web[12].

In the area of Life Sciences, linked data belong to a large number of previously existing databases. However, there is a tendency in this domain to group data from multiple databases into a single repository to offer efficient access to the linked data. Among these services Bio2RDF stands out, it has normalized URIs thereby allowing access to numerous databases of interconnected data of life sciences in RDFs formats. Moreover, this server is compatible with other semantic tools such as browsers Piggy [13] and Tabulator. Another server of linked data is WiWiss, which has brought together information from various databases such as DailyMed[2]. Life Linked Data (LLD) is a platform that enables integration of information by converting it to RDF. LLD represents a semantic repository called OWLIN [14].

3 SBMM: Ontology Based Data Integration Applied to Life Sciences

The System Biology Metabolic Modeling Assistant[3] is a tool developed to search, visualise, manage and annotate both identity data and kinetic data. The possible

---

[1] http://www.insilicodiscovery.com/
[2] http://dailymed.nlm.nih.gov/
[3] http://sbmm.uma.es

inputs for searching metabolic pathways are the pathway's name or the KEGG (Kyoto Encyclopedia of Genes and Genomes) [15] code, a set of enzymes or a correctly annotated model (user-defined or not). Users can retrieve online data on metabolic pathways and then edit them. This tool allows users to export the metabolic pathway to SBML format, enriched automatically without any previous configuration using MIRIAM (Minimum Information Requested In the Annotation of biochemical Models) [16] and CellDesigner 4.0 annotations[4] (this provides a good presentation of the data based on the Systems Biology Graphical Notation, abbreviated as SBGN annotations [17], which are visual notations for network diagrams in systems biology.

In SBMM we have chosen XML, in a broad sense: XML, XML-Schema and XQuery, as the common data model. However, it is also possible to deal with RDF (RDFSchema and SPARQL). This approach is based on Data Services requiring the development of a wrapper. Irrespective of the development process, Data Services are distributed software applications that receive queries in XQuery/SPARQL and return XML/RDF documents. The integration process implies finding a set of mappings between one or several ontologies and the data service schema (expressed as an XMLSchema/RDFS document). These mappings will be the key elements to integrate all the data sources, and they will be the way resource semantics are made explicit. As the proposal is to use ontologies to integrate data, we have chosen a GAV(Global as View) approach. In GAV, each source is related to the global schema (ontology in our case) by means of mappings. Moreover, the use of ontologies will allow us to take advantage of reasoning mechanisms to improve query rewriting.

Thus, the emergence of proposals to carry out this integration process through RDF is completely in line with SBMM and SPA approaches. But, now that the heterogeneity problem (at a syntactic level) has been solved, the Life Sciences domain, heterogeneity at a semantic level must be taken into account (different naming conventions, usage of abbreviations, etc.).

4 Bioqueries: Data Query Design for Life Sciences

Bioqueries[5] is an application for end-users with a biological and bioinformatics profile that aims to start the process towards consuming biological Linked Data through the dissemination of the technology by means of online social networks. Thus, communities would be built around a common interest in certain biological domains to take advantage of public data (currently available in diverse web portals). This will be achieved by sharing a virtual space in a wiki-based portal for the design of SPARQL queries that can be executed in the same environment and documented using natural language descriptions. In order to give the portal an initial critical mass of content to create an active community, it has been populated with a set of around seventy five queries of several biological SPARQL Endpoints. Additionally, any user can add new repositories and queries to any Linked Data

---

[4] http://www.celldesigner.org/
[5] http://bioqueries.uma.es

repository. These queries can be private or publicly shared with other users, the documentation being (detailed descriptions and contextualization) the key to enable other users to understand and use designed queries. This didactic project will join two separate domains, biology and bioinformatics, to obtain specific answers on information from the Linked Data cloud and the approximation of this emerging technology.

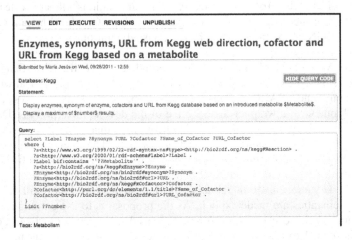

Fig. 1 Visualization of a query. Users can see natural language descriptions as well as the SPARQL query.

The user interface provides the execution of queries directly from the wiki pages. The answer obtained by the service is shown in the wiki interface as HTML, RDF or N-triples formats based on the selected format. For example, a user can access a query related to metabolism (using the Kegg database [15]). This query (Figure 1) displays information about all chemical reactions, enzymes and synonyms of enzymes, cofactors and URL from the Kegg database based on the metabolite introduced.

5 BioSStore: Application Integration through Web Services

Finally, the access to large amounts of data is not sufficient in this domain, and so these data should be available for analysis processes. In this sense, bioinformatics tools have faced the problem of solving the most relevant analysis tasks over biological data. These tools are usually available as Web Services. So, the effort of bringing up all the data together can be made with the use of these services.

The effort to produce standards for the semantic annotation of Web services has guided a lot of the research dealing with the problem of Web services discovery and composition using semantic technologies. However, these studies lack real life large scale scenarios to test their algorithms. In BioSStore[6], we aim to combine the

---

[6] http://biosstore.uma.es

efforts in semantic Web technologies with the huge amount of syntactic (and real) Web services available in bioinformatics. The idea is to enable the semantic annotation of biological Web services (existing and new). This approach will facilitate the service discovery and composition. However, existing tools working with Web services will keep working as the original Web services will not change at the lower level. This Section will show how the registry has been built. The general steps to perform the service annotation process are divided into four main tasks:

1. Information retrieval: the targeting repository is analysed and accessed in order to locate all the information about services (WSDL files, metadata, etc.). The access method will depend on the repository implementation, but available APIs should be used if possible. For populating the repository BioMOBY services[7]are accessed to retrieve the information on biological services.

2. Adaptation phase: This stage includes the transformation of WSDL descriptions to the WSDL 2.0 specification. Available metadata is also analysed and transformed into a common format following an entity-relationship model.

3. Mapping generation: if metadata exists, it will be used to automate the annotation process. Otherwise, the correspondence between the syntactic and the semantic level will have to be done by hand. However, regardless of the specific mechanism to generate the mappings, this phase must follow a quality-effort trade-off as the accuracy of the results relies on it. This means that even when metadata is available a data curation could be useful here to improve the quality of service annotations.

4. RDF transformation: This stage includes the generation of RDF from the annotations created in the previous phase. In this way, enriched service descriptions will be used through a SPARQL API for discovery and composition purposes.

End users will use a Web client which has been developed to provide a powerful and easy to use interface to allow users to locate services. The user interface allows users to make queries by indicating a set of keywords, which will be matched with the service semantic annotations. In this sense, the user interface tends to provide an easy to use interface for biologists and bioinformaticians. Namely, when a user introduces a word, the user interface generates similar terms of the ontology to be selected by the user. For example the user can introduce "NucleotideSequence" to locate services that receive or produce a nucleotide sequence, and this will return a list of services that match this search.

Reasoning in the ontology used to annotate the services is based on the facts introduced into the repository. BioSStore Client User Interface takes advantage of reasoning by providing suggestions in the search filtering for the users, improving their search experience. For example, when the user selects a service type, the system can infer the possible input or output data types, and allows the user to select the one that best fits with his/her enquiry. The use of a reasoning service allows

---

the interface to locate services under the ontology perspective. For example, if we search for services with a "GenericSequence" as input, most of the systems only retrieve the services annotated with this information. By contrast, BioSStore will be able to retrieve services annotated with this input (e.g. "fromGenericToAminoAcidSequence") and those annotated with any of its descendants (e.g. those with an input annotated as "AminoacidSequence", such as "ConvertAAtoFAS-TAAAService").The list of results can be reduced via an additional filtering by restricting the input type, output type or service type.

6 Discussion and Ongoing Work

In this paper we have presented SBMM, a system for using Ontology-based data integration for the development of a useful application for exploring metabolic pathways. So, Bioqueries describes how to provide an environment to show biologists the potential of the emerging technology of Linked Data in Life Sciences. Finally, BioSStore shows how by annotating Web Services in terms of an ontology, the discovery process can be improved.

However, SBMM is nothing new with respect to data integration, Bioqueries provides a way of querying single Linked Data repositories and BioSStore does not take advantage of Linked Data sources. The aim of our ongoing work is to combine these efforts in a new tool for the management of metabolic pathway information. In this sense, several problems are being studied:

- Schema Integration: the different Linked Data repositories follow different schemas for representing the data in RDF. Thus, ontology matching approaches have to be applied. However, some of the used ontologies are large ontologies that cannot be managed. Another problem is the biological differences of concepts between ontologies that make the process of mapping between them difficult. So, novel approaches in this problem have to be taken into account.
- Entity resolution: the Life Sciences is a domain in which databases have grown independently, and so each database uses different naming conventions. Thus, entity resolution approaches should be applied to discover how entities of each Linked Data repository relate.
- Federated Queries: the evaluation of federated queries for Linked Data is a key research field, where traditional approaches can be applied but data distribution and loosely coupled schemas are problems to be taken into account. The optimization of these federated queries is another interesting problem as Linked Data repositories can have large time responses.
- Services for analyzing data: Web Services for analyzing biological data are designed to deal with certain formats. Thus, some developments need to be undertaken to enable these services to automatically translate RDF graphs to the formats required by these services.

- Visualization: the integration of the metabolic pathway data is not the only problem in this domain. The reconciliation of relevant data will end with large sets of data that should be visualized. However, current automatic visualization techniques for metabolic pathways are unable to produce comprehensible graphs due to the large amount of nodes and edges.

7 Conclusions

This paper presents three approaches and a discussion on how to combine them in an approach to deal with metabolic pathway data available as Linked Data. Thus, the effort made in the beginning was focused on bringing semantics to heterogeneous data sets in the Life Sciences (SBMM). Then, the need to solve the problem of consuming lined data (which solves the syntactic problems of the data integration) is dealt with in Bioqueries. Finally, the access to the data in a domain like Life Sciences is just the beginning of the story, and the combination with tools for analyzing these data is needed to extract the relevant knowledge for scientists.

References

[1] Belleau, F., Nolin, M.A., Tourigny, N., Rigault, P., Morisette, J.: Bio2RDF: Towards a Mushup to Build Bioinformatics Knowledge Systems. J. Biomed. Inform. 41, 706–717 (2008), doi:10.1016/j.jbi.2008.03.004

[2] Momtchev, V., Peychev, D., Primov, T., Georgiev, G.: Expanding the Pathway and Interaction Knowledge in Linked Life Data. In: 8th Proceedings of International Semantic Web Challenge, October 25-29. EEUU, Washington (2008)

[3] Risch, T., Josifovski, V.: Distibuted Data Integration through Object-Oriented Mediator Servers. Concurrency and Computation: Practice and Experience 13, 933–953 (2001), doi:10.1002/cpe.607

[4] Tomasic, A., Amouroux, R., Bonnet, P., Kapitskaia, O., Naacke, H., Raschid, L.: The Distributed Information Search Component (Disco) and the World Wide Web. In: Proceedings of the 1997 ACM SIGMOD, pp. 546–548. ACM press (1997), doi:10.1.1.49.5405

[5] Hammer, J., McHugh, J., Garcia-Molina, H.: Semistructured Data: The Tsimmis Experience. In: First East-European Workshop on Advances in Databases and Information Systems-ADBIS, St. Petersburg, Russia (1997), doi:10.1.1.136.465

[6] Roth, T., Arya, M., Haas, L.M., Carey, M.J., Cody, W.F., Fagin, Schwarz, P.M., Thomas II, J., Wimmers, E.L.: The Garlic Project. In: SIGMOD Conference, p. 557 (1996)

[7] Ksiezyk, T., Martin, G., Jia, Q.: InfoSleuth: Agent-Based System for Data Integration and Analysis. In: 25th Annual International Computer Software and Applications Conference, COMPSAC 2001, pp. 474–476 (2001), doi:10.1109/CMPSAC.2001.960655

[8] Bergamaschi, S., Castano, S., Vincini, M., Beneventano, D.: Semantic integration of heterogeneous information sources. Data Knowl. Eng. 36, 215–224 (2001)

[9]   Bornhövd, C.: Semantic Metadata for the Integration of Web-Based Data for Electronic Commerce. In: International Conference on Advance Issues of E-Commerce and Web-Based Information Systems, WECWIS, pp. 137–145 (1999), doi:doi=10.1.1.24.9229

[10] Mena, E., Kashyap, V., Sheth, A.P., Illarramendi, A.: OBSERVER: An Approach for Query Processing in Global Information Systems based on Interoperation across Pre-existing Ontologies. Distributed and Parallel Databases 8, 14–25 (1996), doi:10.1109/COOPIS.1996.554955

[11] Cheung, K.H., Frost, H.R., Marshall, M.S., Prud'hommeaux, E., Samwald, M., Zhao, J., Paschke, A.: A journey to Semantic Web query federation in the life sciences. BMC Bioinformatics 10, 10 (2009), doi:10.1186/1471-2105-10-S10-S10

[12] Langegger, A.: Virtual data integration on the web: novel methods for accessing heterogeneous and distributed data with rich semantics. In: Proceedings of the 10th International Conference on Information Integration and Web-based Applications & Services, pp. 559–562. ACM, New York (2008), doi:10.1145/1497308.1497410

[13] Huynh, D., Mazzocchi, S., Karger, D.R.: Piggy bank: Experience the Semantic Web inside your web browser. J. Web Sem. 5, 16–27 (2007), doi:10.1007/11574620_31

[14] Bishop, B., Kiryakov, A., Ognyanoff, D., Peikov, I., Tashev, Z., Velkov, R.: OWLIM: A family of scalable semantic repositories. Semantic Web 2, 33–42 (2011)

[15] Kanehisa, M., Goto, S., Kawashima, S., Okuno, Y., Hattori, M.: The KEGG Resource for Deciphering the Genome. Nucleic Acids Res. 32, 277–280 (2004), doi:10.1093/nar/gkh063

[16] Le Novère, N., Finney, A., Hucka, M., Bhalla, U.S., Champagne, F., Collado-Vides, F., Crampin, E.J., Halstead, M., Klipp, E., Mendes, P., Nielsen, P., Sauro, H., Shapiro, B., Snoep, L.J., Spence, H.D., Wanner, B.L.: Minimum information requested in the annotation of biochemical models (MIRIAM). Nat. Biotechnol. 23, 1509–1515 (2005)

[17] Le Novère, N., Hucka, M., Mi, H., Moodie, S., Schreiber, F., Sorokin, A., Demir, E., Wegner, K., Aladjem, M.I., Wimalaratne, S.M., Bergman, F.T., Gauges, R., Ghazal, P., Kawaji, H., Li, L., Matsuoka, Y., Villéger, A., Boyd, S.E., Calzone, L., Courtot, M., Dogrusoz, U., Freeman, T.C., Funahashi, A., Ghosh, S., Jouraku, A., Kim, S., Kolpakov, F., Luna, A., Sahle, S., Schmidt, E., Watterson, S., Wu, G., Goryanin, I., Kell, D.B., Sander, C., Sauro, H., Snoep, J.L., Kohn, K., Kitano, H.: The Systems Biology Graphical Notation. Nat. Biotechnol. 27, 735–741 (2009)

Analysing the Effectiveness of Crawlers on the Client-Side Hidden Web

Víctor M. Prieto, Manuel Álvarez, Rafael López-García, and Fidel Cacheda

Abstract. The main goal of this study is to present a scale that classifies crawling systems according to their effectiveness in traversing the "client-side" Hidden Web. To that end, we accomplish several tasks. First, we perform a thorough analysis of the different client-side technologies and the main features of the Web 2.0 pages in order to determine the initial levels of the aforementioned scale. Second, we submit a Web site whose purpose is to check what crawlers are capable of dealing with those technologies and features. Third, we propose several methods to evaluate the performance of the crawlers in the Web site and to classify them according to the levels of the scale. Fourth, we show the results of applying those methods to some OpenSource and commercial crawlers, as well as to the robots of the main Web search engines.

1 Introduction

The World Wide Web (WWW) is currently the biggest information repository ever built. There are huge quantities of information that is publicly accessible, but as important as the information itself is being able to manage it to find, retrieve and gather the most relevant data according to users' needs in every moment.

The programs that process the Web in order to achieve that goal are called crawlers. A crawler traverses the Web following the URLs it discovers in a certain order and analyses the content of each document to obtain new URLs that will be processed later.

From their origins, crawling systems have had to face a lot of difficulties when they access human-oriented Web sites because some technologies are very hard to

Víctor M. Prieto · Manuel Álvarez · Rafael López-García · Fidel Cacheda
Department of Information and Communication Technologies, University of A Coruña,
Campus de Elviña s/n. 15071 A Coruña, Spain
e-mail: {victor.prieto,manuel.alvarez}@udc.es
　　　{rafael.lopez,fidel.cacheda}@udc.es

J.M.C. Rodríguez et al. (Eds.): Trends in PAAMS, AISC 157, pp. 141–148.

analyse (navigation through pop-up menus, different data layers that hide out or appear depending on users' actions, redirection techniques, etc.). The pages that can only be accessed through these technologies constitute what we call Hidden Web [3] and, particularly, the set of pages that are "hidden" behind client-side technologies are called the "client-side Hidden Web".

The objective of this paper is to present a scale for classifying crawlers according to their treatment of the client-side Hidden Web. To that end, we have accomplished some tasks. First, we have analysed the most important client-side technologies, such as JavaScript, VBScript, AJAX, and Flash, as well as some techniques that are often used for illicit purposes, like Redirection Spam [4] and Cloacking [16]. Once those technologies have been analysed, we have enumerated the difficulties that crawlers can find during their traversal. We have also developed a Web site that generates links dynamically and in accordance with the enumerated difficulties in order to know how existing crawlers behave towards them. For this task, we took into account the crawlers of the main Web search engines, as well as for some OpenSource crawlers [6] [9] and some others with a commercial licence. Then, we have discussed some methods to assess the effectiveness of the crawlers, allowing us to choose the appropriate levels in the scale for them.

The structure of this article is as follows. Section 2 discusses the related works. Section 3 introduces the client-side technologies and how they are used to build dynamic Web sites and defines the scale proposed for classifying crawling systems according to their effectiveness in traversing the client-side Hidden Web. Section 4 shows the experimental results performed with open-source and commercial crawlers. Finally, in Sections 5 and 6 we comment the conclusions we have reached and some possible future works.

2   Related Work

There are many studies about the size of the Web and the characterization of its content. However, there are not so many studies about classifying Web pages taking into account the difficulty for crawlers to process their content. According to the data submitted in *W3techs*[1] and *BuiltWith*[2] currently the 90% of the Web pages use JavaScript. In 2006, M. Weideman y F. Schwenke [15] published a study analysing the importance of JavaScript in the visibility of a Web site, concluding that most of the crawlers do not deal with it appropriately.

From the point of view of crawling systems, there are many works oriented to create programs that are capable of traversing the Hidden Web [3]. Server-side Hidden Web crawlers deal with a large quantity of Web sites whose content is accessed by means of forms. This kind of content is copious and has excellent quality. There are some researches that tackle the challenges established by the server-side Hidden Web. We highlight HiWE [11] because it is one of the pioneer systems. Google [7] also submitted the techniques that they use to access information through forms.

---

[1] http://w3techs.com/
[2] http://builtwith.com/

Álvarez *et al.* show DeepBot [2], a prototype of hidden-web crawler able to access hidden content, identifying automatically the web forms and learning to execute queries on them.

Regarding the client-side Hidden Web, there are less studies. From of our knowledge, this is because the crawlers of major search engines have agreements with companies to access data from their servers. However, they only have agreements with major companies in each sector, and a normal crawler has no such agreements.

However, there have been several studies that analyze the issue. Álvarez *et al.* [1] propose the usage of mini-browsers to execute the client-side technologies and so, have access to the hidden content. In 2008, Mesbah *et al.* show a study [8] about the usage of AJAX on the Web, and as can be processed to have access to the data.

On the other hand, as client-side technologies can be used to "deceive" crawling systems, there are some works about detection of what is known as Web Spam (Cloacking [16] [18] [17] [5] or Redirection Spam [4] [5]).

Nevertheless, there is not any scale that allows researchers to classify the effectiveness of the crawling systems according to their level of treatment of client-side Hidden Web technologies.

3   The Scale for Web Crawlers

The client-side technologies are normally used to improve the users' experience, generating content and links dynamically according to users' actions. Among the most used are: JavaScript; AJAX; Flash;Applet and VBScript. In order to create the scale, we have analysed how designers use the aforementioned technologies to build dynamic sites. The following features have been identified:

- Text links, which constitute the lowest level of the scale.
- Simple navigations, generated with JavaScript, VBScript or ActionScript. This includes links that are generated by means of "document.write()" or similar functions, which allows designers to add new links to the HTML code dynamically.
- Navigations generated with an Applet. We divide them in two types: those which are generated from a URL that is passed to the Applet as an argument and those whose URL is created as a string into the compiled code.
- Navigations generated by means of AJAX.
- Pop-up menus, generated by a script code that is associated to any event.
- Navigations generated with Flash. There are two kinds: those which receive the URL as an argument from the HTML code and those which define it inside the ActionScript code.
- Links that are defined as strings in .java files, .class files or any other kinds of text and binary files.
- Navigations generated in script functions. The script can be embedded inside the HTML or it can be located inside an external file.
- Navigations generated by means of several kinds of redirections: a) those specified in the <meta> tag; b) generated by the onLoad event of the <body> tag;

c)generated by a script when an event in other page (e.g.: the onClick event) is fired; d) embedded in script blocks; e) executed in an applet f) Fash redirections.

In addition, the navigations that are generated with any of the identified methods can create absolute or relative URL addresses. For the addresses that are built with scripting languages, it is possible to recognize the following construction methods: a) a static string inside the script; b) a string concatenation; c) execution of a function that builds the URL in several steps.

On the other hand, the different methods we enumerated before can be combined. For example, some Web sites build pop-up menus dynamically, by means of "document.write()" functions. The number of possibilities is unapproachable. Hence, this study only takes into account a reduced but significative subset. It consists of 70 "scenarios" (see "Description" and "Tested Scenarios" columns in Figure 1) that represent the basic types from which the rest of the cases could be obtained by means of combinations. The number of scenarios taken into account could be higher, but it would not get more information about the use of client-side technologies in the Web or about the methods that crawlers use to discover links. For instance, it is not necessary to check the combination of menus and "document.write()" because we can deduce the result from the two base cases that were included separately.

Starting from the aforementioned 70 scenarios, we proposed an initial grouping based on the technologies, the methods for building strings, the location of the code and if the URLs are absolute or relative. This way, we have grouped the scenarios that presented a similar difficulty for crawlers and we have also sorted them by complexity. Figure 1 shows the 8 step scale. The steps represent the capacity to treat the client-side Hidden Web from lower to higher level of complexity.

Level	Description	Tested Scenarios
1	Text Link	1,2
2	JavaScript/Document.Write/Menu – Static String – Embedded	3, 4, 15, 16, 27, 28
2	JavaScript – Concatenated String – Embedded	6
3	Redirect HTML/onBody/JavaScript	51, 52, 53, 54, 55, 56
3	JavaScript con # – Static String – Embedded	63,64
3	VbScript – Static String – Embedded	67,68
4	VbScript – Special Function – Embedded	70
4	JavaScript/Document.Write – Static String – External/Embedded	9, 10, 18
5	Document.Write/Menu – Static String – External	21, 22, 33, 34
5	Menu – Concatenated String – Embedded	30
6	JavaScript/Document.Write/Menu – Concatenated String – External	12, 24, 36
6	Applet – Static String in HTML	43,44
7	JavaScript – Concatenated String – Embedded – Relative	5
7	JavaScript – Special Function – Embedded	7,8
7	JavaScript – Concatenated String – External – Relative	11
7	JavaScript – Special Function – External – Relative	13,14
7	Document.Write – Concatenated String – Embedded/external – Relative	17,23
7	Document.Write – Special Function – Embedded/External	19, 20, 25, 26
7	Menu – Concatenated String – Embedded/external – Relative	29,35
7	Menu – Special Function – Embedded/External	31, 32, 37, 38
7	Link in .class	41,42
7	Ajax Link – Absolute	62
7	Link in .java	39,4
8	Applet – Static String in .class	45,46
8	Flash – Static String in HTML/SWF	47, 48, 49, 50
8	Redirect Applet/Flash	57, 58, 59, 60
8	Ajax Link – Relative	61
8	JavaScript with # – Special Function – Embedded	65,66
8	VbScript – Special Function – Embedded	69

Fig. 1 Link classification by difficulty

Once the scale has been defined, in order to classify the different crawling systems according to the level of complexity of the "links" they process, we propose the following evaluation methods:

- Simple Average: it treats all the scenarios in the same way, without taking into account their difficulty. It shows the crawlers which treat the highest number of scenarios, so they pay more attention to the Hidden Web in general.
- Maximum Level: this model sorts crawlers according to the highest level of difficulty they can process. A crawler obtains a score i if it has the capacity of processing the scenarios of that level and the levels below. There are some crawlers that process a certain level, but they cannot obtain pages from scenarios of lower level. This could be due to some problems like the low PageRank of a Web page and so on. However, this evaluation method assumes that if a crawler is capable of dealing with a level i, it should be able to deal with lower ones.
- Weighted Average: each scenario is assigned a value between 0 and 1, which depends on the number of crawlers that have been able to process it (0 if every crawler has been able to deal with it). This method shows what crawlers can obtain the highest number of difficult resources in the client-side Hidden Web, or resources that most crawlers do not reach.
- Eight Levels: in this model each level has a value of one point. If a crawler processes all the scenarios of one level it obtains that point. For every scenario that the crawler processes successfully, it gets $1/n$ points, where n is the total number of scenarios that were defined for that level.

4 Experimental Results

In order to check how crawling systems deal with the different scenarios, and for ranking them using the scale defined, we created a web site for performing experiments. The "jstestingsite"[3] web site contains 70 links, one for each scenario defined in the scale. We employed it to test the crawlers of the main Web search engines (Google, Bing, Yahoo!, PicSearch and Gigablast) and other OpenSource and commercial crawlers (Nutch [6], Heritrix [9], Pavuk [10], WebHTTrack, Teleport [12], Web2Disk [14], WebCopierPro [13]).

4.1 Summary Results

The left side of Figure 2 shows the results obtained for OpenSource and commercial crawlers. The crawler that achieves the best results is WebCopierPro, which processed 57,14% of the levels, followed by Heritrix with 47,14% and Web2Disk with 34,29%. Only a few get values beyond 25% in most of the levels. It is also important to notice the poor results that they obtained for redirections, especially in the case of WebCopierPro which was not able to deal with any of them, although it gets results of 100% in harder levels. None of the crawlers reached the 100% in

---

[3] http://www.tic.udc.es/~mad/resources/projects/jstestingsite/

	Heritrix	Nutch	Pavuk	Teleport	Web2Disk	WebCopierPro	WebHTTrack	Google	Yahoo
Text Link	2 – 100%	2 – 100%	2 – 100%	2 – 100%	2 – 100%	2 – 100%	2 – 100%	2 – 100%	1 – 50%
Href="JavaScript Link	6 – 50%	3 – 25%	0 – 0%	4 – 33%	5 – 42%	12 – 100%	3 – 25%	6 – 50%	0 – 0%
Document.write Link	6 – 50%	3 – 25%	0 – 0%	3 – 25%	4 – 33%	12 – 100%	2 – 17%	6 – 50%	0 – 0%
Menu Link	6 – 50%	3 – 25%	0 – 0%	3 – 25%	4 – 33%	12 – 100%	2 – 17%	6 – 50%	0 – 0%
Flash Link	0 – 0%	0 – 0%	0 – 0%	0 – 0%	0 – 0%	0 – 0%	2 – 50%	0 – 0%	0 – 0%
Applet Link	2 – 50%	0 – 0%	0 – 0%	0 – 0%	0 – 0%	0 – 0%	2 – 50%	0 – 0%	0 – 0%
Redirects	6 – 60%	6 – 60%	2 – 20%	6 – 60%	4 – 40%	0 – 0%	6 – 60%	6 – 60%	2 – 20%
Class or Java Links	0 – 0%	0 – 0%	0 – 0%	0 – 0%	0 – 0%	2 – 50%	0 – 0%	0 – 0%	0 – 0%
Ajax Link	0 – 0%	1 – 50%	0 – 0%	0 – 0%	0 – 0%	0 – 0%	0 – 0%	0 – 0%	0 – 0%
Links with #	2 – 50%	2 – 50%	0 – 0%	3 – 75%	2 – 50%	0 – 0%	2 – 50%	4 – 100%	0 – 0%
VbScript Link	3 – 75%	3 – 75%	0 – 0%	3 – 75%	3 – 75%	0 – 0%	2 – 50%	1 – 25%	0 – 0%
Static String Link	26 – 62%	19 – 45%	4 – 10%	18 – 43%	22 – 52%	16 – 38%	22 – 52%	17 – 40%	3 – 7%
Concatenated String Link	6 – 50%	2 – 16%	0 – 0%	1 – 8%	1 – 8%	12 – 100%	1 – 8%	6 – 50%	0 – 0%
Special Function String Link	1 – 6%	2 – 13%	0 – 0%	5 – 31%	1 – 6%	12 – 75%	0 – 0%	8 – 50%	0 – 0%
Tests passed	33 – 47%	23 – 33%	4 – 6%	24 – 34%	24 – 34%	40 – 57%	23 – 32%	31 – 44%	3 – 4%

Fig. 2 Summary of results of OpenSource and Commercial crawlers (left side) and the crawlers of the main search engines (right side)

redirections. This happens because none of them has been able to process pages with redirections embedded in Applets or Flash, since these technologies were not executed.

The right side of Figure 2 contains the results obtained for the main Web search engines. It does not show the results for Bing, Gigablast and PicSearch, since they have not indexed the testing Web site. Only Google and Yahoo! have indexed it. Google has processed 44.29% of the links. GoogleBot has processed 50% of many of the proposed levels. The other half has not been processed because the crawler has not analysed some external files. If it was the case, GoogleBot would achieve much better results. The links that GoogleBot has not processed included technologies like Flash, Applets and AJAX or files like .class and .java.

Regarding the types of links that in general were processed by the crawlers analyzed, we found that only 42.52% of statics links were retrieved, 27.38% of links generated by concatenation of strings and 15.18% for links that were generated by functions. These results show that the crawlers try to discover new URLs by processing the code as text, using regular expressions, instead of using scripting interpreters and/or decompilers. So, they cannot extract those links which were generated by a complex method.

4.2   Ranking the Crawlers According to the Scale

Figure 3 shows the result of classifying the crawling systems according to the scale and the assessment levels we had proposed.

We can see that WebCopier, Heritrix and Google get the best results in the Simple Average method. For Maximum Level, Google places first since it processes level 8 links. It is followed by WebCopier (7 points) and Heritrix (6 points). As Google achieves the maximum level in this model but not in others, we can conclude that it does not try some scenarios because of its internal policy. Once again, WebCopier, Google and Heritrix have obtained the best results in the Weighted Average model. Very similar results have been obtained in the Eight Levels method. This means that the three top crawlers have dealt with a big quantity of levels in each group or

	Heritrix	Nutch	Pavuk	Teleport	Web2Disk	WebCopier	WebHTTRack	Google	Yahoo
Simple Average	3,77	2,63	0,46	2,29	2,74	4,57	2,40	3,54	0,34
Maximum Level	6,00	7,00	3,00	4,00	5,00	7,00	6,00	8,00	3,00
Weighted Average	1,63	0,99	0,14	0,75	1,03	3,42	0,86	2,34	0,11
Eight levels	6,00	4,30	1,20	4,00	4,55	4,55	3,40	3,70	0,70

Fig. 3 Results according to the proposed scales

they have gone through links that were part of a group with few links, which makes each link more valuable. We conclude that the best crawlers in both quantity and quality are Google and WebCopier, followed by Heritrix, Nutch and Web2Disk. It is important to highlight the results of GoogleBot. Although it is oriented to traverse all the Web and it has a lot of performance and security requisites, it takes into account a wide range of technologies.

5 Conclusions

This article proposes a scale that allows us to classify crawling systems according to their effectiveness accessing the client-side Hidden Web. In order to classify the different crawling systems, we have created a Web site implementing all the difficulties that we had included in the scale.

Analyzing the results we can make the following recommendations about creating a web page: use JavaScript, embed JavaScript code in HTML, avoid dynamic creation of URLs and use HTTP or HTML code for the redirects.

The best crawlers are Google and WebCopier, followed by Heritrix, Nutch and Web2Disk. We can say that most of the times they try to discover new URLs processing the code as text, using regular expressions. This allows them to discover a big amount of scenarios. In addition, the major crawlers have agreements with companies in each sector, allowing them direct access to data. With these policies, the crawlers can treat directly the data and save resources. However, only the major search engines have these agreements. We conclude that in present, most of the URLs which are located in the client-side technologies are not discovered.

6 Future Work

Among the studies that we propose as a continuation of this work, we have the improvement of the assessment methods in order to take into account the frequency of use of each technology on the Web. This will allow us to know the volume of information that is beyond the scope of the crawlers, since nowadays they do not treat all the scenarios. We can use this information to know the convenience of analysing the client-side Hidden Web. We also are interested in in studying how the features (the topics, the number of visits, etc.) of a Web site can affect the crawling process and the indexation of its pages.

Acknowledgements. This work was partly supported by the Spanish government, under projects TIN 2009-14203 and TIN 2010-09988-E.

References

1. Álvarez, M., Pan, A., Raposo, J., Hidalgo, J.: Crawling Web Pages with Support for Client-Side Dynamism (2006)
2. Álvarez, M., Raposo, J., Pan, A., Cacheda, F., Bellas, F., Carneiro, V.: Crawling the Content Hidden Behind Web Forms. In: Gervasi, O., Gavrilova, M.L. (eds.) ICCSA 2007, Part II. LNCS, vol. 4706, pp. 322–333. Springer, Heidelberg (2007)
3. Bergman, M.K.: The deep web: Surfacing hidden value (2000)
4. Chellapilla, K., Maykov, A.: A taxonomy of javascript redirection spam. In: Workshop on Adversarial Information Retrieval on the Web, AIRWeb 2007, pp. 81–88 (2007)
5. Gyongyi, Z., Garcia-Molina, H.: Web spam taxonomy (2005)
6. Khare, R., Cutting, D.: Nutch: A flexible and scalable open-source web search engine. Technical report (2004)
7. Madhavan, J., Ko, D., Kot, L., Ganapathy, V., Rasmussen, A., Halevy, A.: Google's deep web crawl. Proc. VLDB Endow. 1, 1241–1252 (2008)
8. Mesbah, A., Bozdag, E., van Deursen, A.: Crawling ajax by inferring user interface state changes. In: Web Engineering, ICWE 2008, pp. 122–134 (2008)
9. Mohr, G., Kimpton, M., Stack, M., Ranitovic, I.: Introduction to heritrix, an archival quality web crawler. In: 4th International Web Archiving Workshop, IWAW 2004 (2004)
10. Pavuk Web page (2011), http://www.pavuk.org/
11. Raghavan, S., Garcia-Molina, H.: Crawling the hidden web. In: Proceedings of the 27th International Conference on Very Large Data Bases, VLDB 2001, pp. 129–138. Morgan Kaufmann Publishers Inc., San Francisco (2001)
12. Teleport Web page (2011),
    http://www.tenmax.com/teleport/pro/home.html
13. Web Copier Pro Web page,
    http://www.maximumsoft.com/products/wc_pro/overview.html
14. Web2Disk Web page (2011),
    http://www.inspyder.com/products/Web2Disk/Default.aspx
15. Weideman, M., Schwenke, F.: The influence that JavaScript has on the visibility of a Website to search engines - a pilot study. Information Research 11(4) (July 2006)
16. Wu, B., Davison, B.D.: Cloaking and redirection: A preliminary study (2005)
17. Wu, B., Davison, B.D.: Identifying link farm spam pages. In: Proceedings of the 14th International World Wide Web Conference, pp. 820–829. ACM Press (2005)
18. Wu, B., Davison, B.D.: Detecting semantic cloaking on the web. In: Proceedings of the 15th International World Wide Web Conference, pp. 819–828. ACM Press (2006)

Information Extraction Framework

Hassan A. Sleiman and Rafael Corchuelo

Abstract. The literature provides many techniques to infer rules that can be used to configure web information extractors. Unfortunately, these techniques have been developed independently, which makes it very difficult to compare the results: there is not even a collection of datasets on which these techniques can be assessed. Furthermore, there is not a common infrastructure to implement these techniques, which makes implementing them costly. In this paper, we propose a framework that helps software engineers implement their techniques and compare the results. Having such a framework allows comparing techniques side by side and our experiments prove that it helps reduce development costs.

Keywords: Information Extraction Framework Architecture.

1 Introduction

The Web contains a huge amount of information and is a still growing data container. This unlimited repository aroused enterprises' interests in exploiting web information, so new applications that consume and analyse this information have emerged. Applications such as businesses intelligence and other marketing tools need data from the Web to help users making decisions and offer best service. Unfortunately, the information in the Web is embedded in HTML tags and in other contents that in many cases are not interesting. Information extraction is used in these cases to obtain the information in which user is interested and to discard the other [6].

Information extraction from the Web is the task that extracts relevant information from web pages, where relevant is relative to the use case and the user's intentions. During the last decades, many proposals on information extractors have been introduced, but the web has changed and many of these proposals are not useful anymore.

Hassan A. Sleiman · Rafael Corchuelo
University of Sevilla
e-mail: {hassansleiman, corchu}@us.es

J.M.C. Rodríguez et al. (Eds.): Trends in PAAMS, AISC 157, pp. 149–156.
springerlink.com   © Springer-Verlag Berlin Heidelberg 2012

According to a recent report [4], developing and maintaining information extractors is still a tedious process because of the lack of development support tools. Comparing information extractors are usually compared by their concepts, such as the surveys [2] and [11]. Empirical comparisons are still tedious since they require the development of other proposals and need to be performed under coherent conditions.

Relying on a framework in the domain of information extractors has many benefits. Using a framework in developing proposals reduces development and testing costs. If the framework is accompanied by a collection of datasets, then it shall also help compare different techniques homogeneously.

In this proposal, we present an overview of the framework architecture which has been validated by developing several techniques. The time needed to develop some techniques using our framework is compared to the time that was necessary to develop the same techniques without the framework to show the costs reduction. Furthermore, cross validation is used to test proposals to obtain comparable precision and recall between the different techniques.

This paper is organised as follows: First, Section 2 classifies and lists related work briefly and then, Section 3 describes the architecture of the framework. The framework is then used to develop a set of proposals and the experimental results are reported in Section 4. We conclude our work in Section 5.

2 Related Work

The high number of proposals on information extraction makes us classify them according to some criteria to detect the approaches for which our framework shall adapt. Information extractors can be used to extract and structure information from free text web pages, such as news and blogs, or from result and detail web pages, such as web pages with one or more result records and detail web pages with information about a certain product. Our work focuses on the second type of information extractors used for semi-structured web pages.

Information extractors for semi-structured web pages can be classified into two groups: a heuristic based group and a rule based group. The heuristic based group contains proposals that are based on predefined heuristics. Although these heuristics can be seen as rules, the difference between this group and the rule based group is that they are totally independent from the web page on which they are applied. These heuristics are not modifiable and are not inferred by an automatic technique. Techniques like Stavies [13], Alvarez et al. [1], and ViPER [14] are heuristic based.

The rule based group contains information extractors that are configurable by means of rules. Beyond handcrafted information extraction rules, there are many proposals in the literature to learn them in a supervised and in an unsupervised manner. Supervised techniques require user intervention to learn these rules. Generally, it needs the user to annotate a set of web pages by selecting and assigning a type for the relevant information in a set of web pages used then in the learning process. Techniques like Softmealy [7], WIEN [9], Stalker [12], and DEByE [10] are supervised techniques. This latter group also contains a number of unsupervised techniques, which do not requiere the user to provide annotations. Input web pages are analysed to detect repeating patterns or templates used at the server side to generate these pages. Techniques like FiVaTech [8], RoadRunner [5], DEPTA [16], and DeLa [15] belong to this group.

3   Architecture of Our Framework

The framework is composed of six packages. These packages can be used during the rule learning process for the rule based information extractors or in testing extraction results for all the types of information extractors described in the previous section. These packages are Dataset, Annotator, Learners, Tokeniser, Cross Validator, and Utilities, which are explained in the following subsections:

3.1   Datasets

This package, contains all the information annotated by user during the annotation process or extracted by an information extractor during an extraction process, see Figure 1. Package classes are described below:

- **Dataset:** This is a map-like structure that associates a set of annotations to a number of web pages. These annotations represent the relevant data in this web page and are represented by a Resultset.
- **Resultset:** A class that contains the annotations that mark the relevant information in a web page. Each annotation has its description besides the relations between these annotations.
- Locators are pointers to each annotated fragment in a web page. They are of two types: TreeLocators which contain an XPath that points to annotation's node or TextLocators which points to the beginning offset and the length of annotation in the web page.
- Views can be created for a web page: a TextView offers working with the text contained inside a web page and a TreeView which can be used to work with the HTML tree and its nodes.

3.2   Annotator

A tool that helps users download and annotate web pages to create Datasets. First, the user shall select an ontology which is used to assign a type and a relation between the annotations. Then, he or she navigate to web pages and add them to the created Dataset. Once added, contents from this web page can be selected, dragged and dropped into the ontology. This allows the creation of individuals of a certain class and assigning properties to them, saving their locators too. The tool also checks and warns for possible errors during the annotation process. Datasets can be then loaded and modified in the tool, or used in the framework for learning and testing tasks.

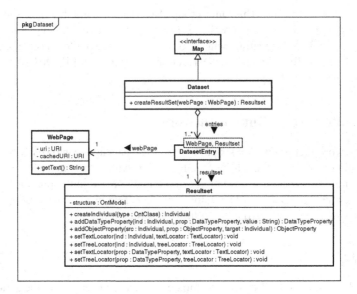

Fig. 1 Dataset Package

3.3 Learners

The Learners package provides an interface implemented by different techniques and other other classes used during rule learning process. Package classes are described briefly next:

- **Learner:** It is an interface that provides a number of template methods software engineers must provide to implement their techniques.
- **SkeletonLearner:** This class transforms a set of annotations into a Transducer in which each state identifies a type of data to be extracted. The transitions between them maintain the separators found in the input web page, which can later be used by the learning algorithms to learn transition rules.
- **Transducer:** A class that models state machines that can be learnt incrementally and executed to extract information from a web page. Thanks to the Skeleton-Learner class, software engineers need only focus on learning the transitions.

3.4 Tokeniser

This is a configurable tokeniser that allows users to define a hierarchy between types of tokens, generalising and specialising them during the learning process, see Figure 2. The main classes of this package are described briefly here:

- **TokeniserConfig:** A class that helps read the XML configuration file where classes and hierarchy are declared and create the token classes. It maintains the defined structure to be used during the learning process.

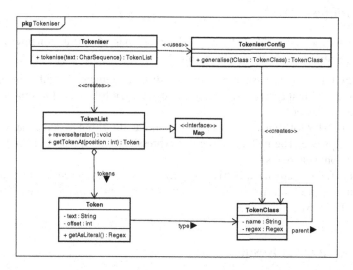

Fig. 2 Tokeniser

- **TokenClass:** This class refers to a type of token. Token classes are defined in the tokeniser's configuration file where a name and a regular expression for each TokenClass is defined.
- **Tokeniser:** A class that tokenises input text by searching for matchings of the defined token classes on this text and returning a TokenList.
- **TokenList:** It is a map that saves the tokens according to their position. It allows to sort and search for specific tokens very efficiently.
- **Token:** This class is created by the tokeniser by assigning a TokenClass to it when it matches the regular expression of a TokenClass. It can be converted into a regular expression in case of specialisation.

3.5 Cross Validator

Our framework provides a cross validation package to evaluate information extractors with the following classes:

- **TestUtilites:** This class is used to compare extraction results with an annotated dataset. It calculates precision, recall, F-measure, accuracy, specificity and sensitivity for each type of data in these datasets.
- **CrossValidation:** Implementes a *K* cross-validator, where k is typically 10.
- **Stats:** This class collects information during the cross validation process. At the end of the cross validation process, this class calculates statistical information about each one of the parameters, e.g., Precision and Recall.

3.6   Utilities

This package includes a set of classes used in more than one extraction rules learning proposal. Some of these utility classes are:

- **StringAligner:** This class implementes a string alignment algorithm inspired by FiVaTech [8] that aligns a set of input sequences of tokens and returns a unique aligned sequence.
- **PatternDetector:** It uses a FiVaTech [8] similar algorithm to detect pattern in an input sequence. The result is a regular expression that represents fixed,repeated and optional elements.
- **PatriciaTree:** It creates a Patricia tree from a set of token sequences and builds a regular expression from this tree. This is used in proposals such as DeLa [15] and IEPAD [3].

4   Experimental Results

To validate our framework, we have implemented a number of proposals in the literature. We provide a toolkit with the following learners NLR, SM and FT which are inspired by [9], [7] and [8] respectively. The time necessary for their development using our framework is compared to the development time that was necessary without using our framework. We have also compared their performance regarding precision and recall on a homogenous collection of datasets.

Table 1 shows the time that was necessary to develop and test the techniques in the first column. The second and the third columns show the time that was necessary to develop these techniques without using the framework and using it. The costs reduction is clear since the framework allowed reusing components and the same datasets were used in every implementation. The last column shows the reduced time percentage, the arithmetic mean of the reduction percentage is 57.51%.

Table 1 Comparing implementation times for NLR, SM, and FT

Technique	Without Framework	Using Framework	Reduced time percentage
NLR	145hrs	32hrs	77.94%
SM	123hrs	87hrs	29.27%
FT	176hrs	61hrs	65.34%

Table 2 reports the results of applying these techniques in practice on several datasets compared side by side. The first column contains the used datasets. Other columns contain precision (P) and recall (R), besides the time that was necessary to learn extraction rules by each technique for each dataset. Each dataset contains 30 web pages, and the results regarding precision and recall were calculated using 10-fold cross validation.

Table 2 Comparing precision and recall of NLR, SM, and FT techniques

Dataset	NLR			SM			FT		
	P	R	T(s)	P	R	T(s)	P	R	T(s)
doctor.webmd.com	0.62	0.62	989.78	0.98	0.95	11.14	0.83	0.61	4.29
extapps.ama-assn.org	0.61	0.61	384.84	0.79	0.38	4.46	0.70	0.58	3.65
www.dentists.com	1.00	1.00	18.82	0.64	0.62	2.32	1.00	0.30	1.84
www.drscore.com	0.80	0.06	14.25	1.00	0.86	4.87	0.81	0.05	3.31
www.steadyhealth.com	0.75	0.72	265.75	1.00	0.96	11.68	1.00	0.78	6.21
classiccarsforsale.co.uk	0.49	0.38	39.14	1.00	0.80	11.89	0.96	0.23	7.62
internetautoguide.com	0.30	0.21	85.23	0.30	0.21	11.68	0.91	0.67	5.87
www.autotrader.com	0.88	0.70	130.15	1.00	0.93	18.34	0.88	0.22	10.59
www.carmax.com	0.84	0.81	39.82	0.99	0.90	8.85	0.89	0.80	5.67
www.carzone.ie	0.84	0.81	37.85	0.99	0.67	6.90	0.98	0.66	4.64

Note that these techniques can obtain better precision and recall by adding more web pages to these datasets, but this is not our case since we are just obtaining comparable results which allows us selecting the extraction rules learning algorithm that best fits the web site we are interested in.

5 Conclusions

In this paper we have described our information extraction framework. We also reported our first experimental results which confirms the fact that using the framework reduces costs and allows side by side comparison providing comparable results. Development costs were reduced 57.51%. Future proposals that use our framework and our datasets can compare their result with the obtained results here without having to implement these techniques again neither annotate the same web pages.

Acknowledgements. Supported by the European Commission (FEDER), the Spanish and the Andalusian R&D&I programmes (grants grants TIN2010-21744-C02-01, TIN2007-64119, P07-TIC-2602, P08-TIC-4100, TIN2008-04718-E, and TIN2010-09988-E).

References

[1] Álvarez, M., et al.: Extracting lists of data records from semi-structured web pages. Data Knowl. Eng. 64(2) (2008)
[2] Chang, C.-H., et al.: A survey of web information extraction systems. IEEE Trans. Knowl. Data Eng. 18(10) (2006)
[3] Chang, C.-H., Lui, S.-C.: IEPAD: information extraction based on pattern discovery. In: WWW (2001)

[4] Chiticariu, L., et al.: Enterprise information extraction: recent developments and open challenges. In: SIGMOD Conference (2010)

[5] Crescenzi, V., et al.: Roadrunner: Towards automatic data extraction from large web sites. In: VLDB (2001)

[6] de Viana, I.F., Hernandez, I., Jiménez, P., Rivero, C.R., Sleiman, H.A.: Integrating Deep-Web Information Sources. In: Demazeau, Y., Dignum, F., Corchado, J.M., Bajo, J., Corchuelo, R., Corchado, E., Fernández-Riverola, F., Julián, V.J., Pawlewski, P., Campbell, A. (eds.) Trends in PAAMS. AISC, vol. 71, pp. 311–320. Springer, Heidelberg (2010)

[7] Hsu, C.-N., Dung, M.-T.: Generating finite-state transducers for semi-structured data extraction from the web. Inf. Syst. 23(8) (1998)

[8] Kayed, M., Chang, C.-H.: FiVaTech: Page-level web data extraction from template pages. IEEE Trans. Knowl. Data Eng. (2010)

[9] Kushmerick, N.: et al. Wrapper induction: Efficiency and expressiveness. Artif. Intell. 118(1-2) (2000)

[10] Laender, A.H.F., et al.: DEByE - data extraction by example. Data Knowl. Eng. 40(2) (2002)

[11] Muslea, I., et al.: Extraction patterns for information extraction tasks: A survey. In: AAAI-1999 Workshop on Machine Learning for IE (1999)

[12] Muslea, I., et al.: Hierarchical wrapper induction for semistructured information sources. Autonomous Agents and Multi-Agent Systems 4(1/2) (2001)

[13] Papadakis, N., et al.: Stavies: A system for information extraction from unknown web data sources through automatic web wrapper generation using clustering techniques. IEEE Trans. Knowl. Data Eng. 17(12) (2005)

[14] Simon, K., Lausen, G.: ViPER: augmenting automatic information extraction with visual perceptions. In: International Conference on Information and Knowledge Management (2005)

[15] Wang, J., Lochovsky, F.H.: Data extraction and label assignment for web databases. In: WWW (2003)

[16] Zhai, Y., Liu, B.: Structured data extraction from the Web based on partial tree alignment. IEEE Trans. Knowl. Data Eng. 18(12) (2006)

Behavior Pattern Simulation of Freelance Marketplace

Vadim Zuravlyov, Anton Matrosov, and Dmitrijs Rutko

Abstract. Labour market is expanding rapidly nowadays. Therefore, recruitment processes, namely, the processes of job searching and job offer have become more complicated and there is a strong tendency to automate and integrate it in special freelance marketplaces, i.e. web solutions which seek for the best match for a buyer and seller analyzing the services and items they offer. The aim of the current research is to find possible improvements to the existing system via the implementation of the Multi-Agent-System paradigm. We model the whole marketplace as a continuous process with different agents (freelancers) and propose several behavior models of agents. We analyze different strategies with the emphasis on agent profit and employer costs and present simulation results which complement our research.

Keywords: freelancers, behavior patterns, Multi Agent Systems, simulation, XML.

1 Introduction

Self-employment has become very popular in recent decades. For instance, according to the latest research about 40% of all IT employees are not employed in a regular employer-employee relationship in Germany [1].

The term "freelancer" is not clearly defined and is usually interpreted as self-employed employee. Different legal and economical definitions could be used in

Vadim Zuravlyov
Faculty of Computer Science, Riga Technical University, Riga, Latvia
e-mail: zuravlov@gmail.com

Anton Matrosov · Dmitrijs Rutko
Faculty of Computing, University of Latvia, Riga, Latvia
e-mail: antmatrik@gmail.com, dim_rut@inbox.lv

J.M.C. Rodríguez et al. (Eds.): Trends in PAAMS, AISC 157, pp. 157–164.
springerlink.com   © Springer-Verlag Berlin Heidelberg 2012

different countries as well. Usually, the following three main areas are studied: media (press, radio or television); consulting companies; IT industry[2-3].

We define the notion of freelancer as a self-employed employee which performs short-term projects of day jobs for any type of work.

Different freelance marketplaces have been studied in various papers [4-6]. The main goal of creating a freelance marketplace is to centralize all the data about project proposals and freelancers. These marketplaces are based on several groups of attributes (more details in [7]) for proposition estimation: pay, working hours, unions (owner, organization etc.).

A limited number of attributes is not sufficient for creating an efficient freelancer resource management system [12]. We propose to extend the existing attribute groups and, therefore, enrich the current behavior patterns.

The proposal is validated by simulating a freelancer marketplace. Multi-Agent Systems (MAS) were used as the main simulation paradigm. Each freelancer acts as an independent agent with unique behavior patterns and its values. Existing freelancer marketplaces and application programming interfaces (APIs) [8] are used in the research. Freelancer privacy policy and security policy are emphasized as well; therefore, the information storage decentralization principle is used.

We introduce an advanced approach for modeling and analyzing freelance marketplace through introducing new attributes which gives us a unique opportunity to make practical experiments and make conclusions based on that experimental data. This allows us to model human behavior and deeply analyze different strategies and incentives in various scenarios.

2   Current Situation and Research Direction

In this study different standard and non-standard terms related to freelancers are used, which are summarized in the list below.

- *Freelancers* – self-employed employees who carry out short-term projects and day jobs. This notion can mean any type of work.
- *Freelancer marketplace* –systems that match freelancers and employers. Service providers, or freelancers, create a profile where they include a description of the services which they offer, examples of their work and in some cases information about their rates. Employers post projects outlining their requirements. Freelancers then will bid for these projects on a fixed price or hourly basis.
- By *attributes* the authors mean properties that describe freelancers, employers or projects. Attributes are always public and usually describe project requirements or freelancer skills.
- *Behavior patterns* imply behavior rules for any freelancer marketplace participant. They are usually private and in contrast to attributes could not be analyzed by other marketplace participants.

Currently every freelancer market participant performs a lot of handwork. Every employer creates project proposals and afterwards monitors the progress of competition - checks freelancers' profiles and finally chooses the winner. Usually freelancers make bids for many projects to win at least one. Freelancers have to track new projects that constantly appear in freelancer market project database, plan participation in these projects and track project progress where the freelancer already won the contest.

This model is relevant for medium and large projects and a relatively small project database. Unfortunately this model does not fit our expectations and assumptions made before. Automation is a very important requirement for small daily jobs and projects database with millions of active project proposals.

3 Data Management based on Properties of Individual Objects

Data-And-Rules-Saved-In-Resource (DARSIR) concept (see more in [10-12]) – is a new data management based on properties of individual objects, which is based on decentralization principles. It means that all properties required for decision-making (attributes and rules) are saved in object (freelancer).

DARSIR concept is based on the following basic components [11]:

- *Resource* (DARSIR resource) – any object of the living or lifeless nature which is involved in the working process of an information system. The information of a concrete resource and its interrelationship with other objects (or types of objects) must be stored in this element.
- *Resource Physical Markup Language* (RPML) – XML-like language that is developed specially for DARSIR concept.

The key idea of DARSIR concept is ensuring privacy and security policy through data decentralization. In previous authors' research [10] this approach was already used. DARSIR concept is suitable for current research also because all needed attribute groups and behavior rules are already defined and implemented in RPML. Unique behavior patterns are also stored in the object through RPML. The actions in resource are done using *triggers*. The trigger (set of actions) is launched after a defined event.

4 Freelancers Resource Management Using MAS

Multi Agent systems (MAS) are the systems composed of multiple interacting computing elements, known as agents. Agents are computer systems, situated in some environment, with two important capabilities. First, they are at least to some extent capable of autonomous action, e.g. deciding for them based on environment information what they need to do in order to satisfy their design objectives. Second, they are able to interact with other agents and changing the environment in some way – not simply by exchanging data, but by engaging in analogues of the kind of social activity that we all are engaged in every day in our lives, i.e. cooperation, coordination, negotiation, and the like [9].

Typically multi-agent systems research refers to software agents. However, the agents in a multi-agent system could equally well be robots humans or human teams. In this study multi agent system with software and human being agents will be studied.

We propose a Multi-Agent system model where each freelancer marketplace participant acts as a separate agent. Freelancers' API is used for the communication between the freelance marketplace the project database, the freelancer and, the employer. Each agent has its own behavior pattern definitions and attributes. The behavior patterns are securely stored on agent's side using DARSIR concept. This means that patterns are not available for other freelance marketplace participants and cannot be sold or used among participants. Depending on these behavior patterns the agent can act fully autonomously without human intervention or semi-automatically involving human for final decision making.

As a result, the freelancer marketplace participant defines its behavior patterns once, instead of permanent candidate searches and project selections. Behavior patterns are not static and will evolve together with the owner. Changes in behavior patterns can be caused by manual intervention of a human or automatically evolve based on previous agent experience.

The task of the freelancer is to choose projects for participation and the task, for employer is to choose the winner for his project. Both tasks can be split into two phases – prioritization phase and decision making phase. First, the freelance marketplace participant prioritizes offers (project offers or bids made by freelancers) and then makes a decision. Therefore, behavior patterns can also be split into two groups. The first group is devoted to offers prioritization and the second group focuses on making a decision.

5 Attributes

One of the key elements of the proposed solution is an attribute. By attributes the authors mean public properties that describe buyers, sellers or projects. It was decided that only freelancer actions will be examined while defining the paper purposes.

Three groups of attributes are described and systematized in different research papers. As the basis the authors took group definitions from the research [7]: pay, working hours (amount of hours for completing the project, deadline) unions (owner, organization etc.).

Automated freelance management system [8] uses the following attributes that will be taken as the basis:

- *Budget* – the sum that the employer is ready to pay for project implementation.
- *Created* – the date of project offer publication.
- *Bidding ends* – the date of the project offer end. The period of offer validity is entered in days, and cannot be more than 60 days.
- *Project Creator* – an employer that orders project implementation.
- *Buyer rating* – employer's rating that is measured using the scale from 0 to 5.
- *Description* – the description of a project from the employer in a free form.
- *Job Type* – describes what exactly should be done.

One of the attributes "Description" is extended into two additional attributes that are required for the automation process. These attributes are:*area* – quantitative parameter that can be used to estimate the amount of work that should be done; *location* – GPS coordinates of the area; an address; a city; a place where the work should be done. This parameter will be used for more precise analysis of freelancer's expenses.

6 The Example

The example of a freelancer project is described in this chapter. Let us assume that there is a freelancer working with lawn mowing. This job usually takes several hours for one project. So there can be several projects in one day. Our task is to help the freelancer in choosing the best projects for him. Ideally, we would like to provide the freelancer with a project timetable for the next week automatically. The main project parameters that the employer is entering in the freelancer marketplace database are pay and working hours.

More detailed information about the project could reduce the project price. For example, if the employer adds the exact location to project attributes describing the project, the freelancer can take this into consideration and if his previous job was close, then the expenses for transfer can be reduced.

Adding more attributes to the project is beneficial for the employer because he can reduce project cost and at the same time is beneficial for the freelancer because he will be able to choose the most profitable project. This hypothesis can be checked by creating a simulation and comparing the simulation results for different strategies.

7 Simulation

To prove the importance of the attribute and check the proposed method we created a simple simulation for the freelancer marketplace. But first, it is required to define the solution comparison method. This will allow to compare the provided optimization methods with each other. There are two key parameters in the freelancer management system: average project costs and average freelancers profit.

The authors propose to use these two parameters for the optimization method comparison. At the current stage of the research the authors simplify the simulation approach to get the first results. In the future the research authors intend to extend the optimization comparison methods significantly by adding more parameters: contentment (employer's and freelancer's), order completion speed, the amount of non-completed task, etc.

The simulation is fully automated. Each agent has its own behavior pattern and fixed bidding selection strategy which complies with DARSIR concept described in part 3. The simulation is based on the rules listed below:

- N agents, M employers, each employer makes K advertisements resulting in Z = M*K (max) advertisements;
- each advertisement has x, y - location coordinates. The location is private, and not publicly available. Each agent has its own private location which is used to calculate the distance to work / the costs to travel per task;
- each agent has its own price for work, which is fixed;
- each agent makes B bids randomly over Z advertisements;
- each agent can perform max W works per day / per round;
- after the bidding round is over each employer selects the cheapest bid;
- make R rounds,calculate the average agent profit / employer payments;
- apply learning strategy - after agents change their strategy (increase or decrease price for their work) depending on the average profit vs. previous strategy.

Next experiments may include different strategies like punishment for not performed work - if the agent makes a bid, the bid is winning, but the agent refuses to perform a task, then we could introduce some incentives to deal with the situation.

8 Simulation Results

The results are shown in the following charts. Fig.1 refers to the first simulation type when the information about the location is private (hidden) e.g. is not publicly open and not available for all agents. Here it is clearly seen that the acceptance rate of the project increases correspondingly with the increasing maximum number of bids. So, the more bids the agent makes, then the chances for project to be accepted by at least on agent increases.

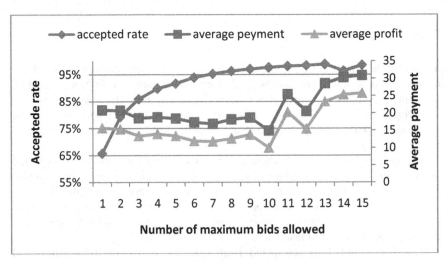

Fig. 1 Accepted rate and average payment depending on number of bids

Another trend line is average payment (for employer) vs. average profit (for agent). These lines go in parallel as there is a fixed gap for average travel expenses. But as it could be seen when the number of bids reaches some threshold (in our case maximum number of bids equals to 10) then agents demand higher prices for their work and project owners (employers) have to pay it.

The following chart (Fig.2) demonstrates the same simulation results with one condition changed, i.e. the geographical information becomes publicly available, so agents can use it when making project proposals (offering work for a specific price).

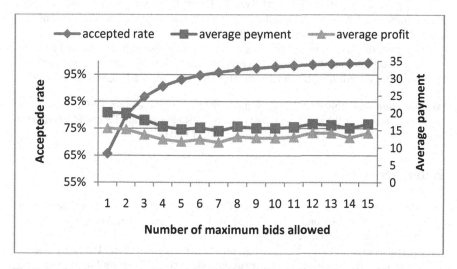

Fig. 2 Accepted rate and average payment depending on number of bids

So agent behavior stays generally the same. The acceptance rate increases with the number of maximum bids allowed growing up. But in this case when we use the information on additional attribute (geo-location), the wages are less, the profit is higher (as there is a smaller gap for travel expenses) and that is rather important payment stabilizes at the certain level and do not go up after some threshold.

9 Conclusions and Future Research

The current structure and the basic aspects of the freelance marketplace have been discussed in this paper. The authors proposed the improvements for the model and presented the main guidelines for introducing additional attributes (pieces of information) to each project. The given information allows considerable improvements in the sense of organizing automatic freelance actions. This concept enables the agents to choose a more appropriate project and make reasonable bids (work prices) for most of the projects. The employers are able to benefit from these techniques as well – this reduces their overall costs.

As the main paradigm for the entire modeling and simulation process Multi Agent System (MAS) was used. This means that each agent behavior is unique as each agent tries to maximize its own outcome. There is no central hub in this approach so freelancer privacy rules are observed.

To provide the empirical results and complement our research with the estimated data we created a simplified freelance marketplace simulation. It defines basic framework for result-based comparison of different models used. In our model we have used two basic strategies:1) the information about the location is private, so the agent cannot access it; 2) the information about the geographical location for the project is available to all marketplace participants.

This gave us a possibility to analyze different behavior patterns and compare the average project costs and employer expenses for each project. The current tool could be used for new models and simulations.

As a part of this study we propose the following research directions:1) implementing additional attributes and different models; 2) analyzing behavior patterns; 3) project organization with priority and applying the proposed ideas in other fields.

Acknowledgments. This work has been supported by the European Social Fund within the projects «Support for the implementation of doctoral studies at Riga Technical University» and "Support for Doctoral Studies at the University of Latvia".

References

1. Hoffmann, E., Walwe, U.: Wandel der Erwerbsformen: Was steckt hinter den Veranderungen? In: IABKompendiumArbeitsmarkt - und Berufsforschung, pp. 135–144. Nurnberg, Germany (2002)
2. Statistisches Bundesam: Bevolkerung und Erwerbstatigkeit 2002. In: Beruf, Ausbildung und Arbeitsbedingungen der Erwerbsta-tigen (Ergebnisse des Mikrozensus), pp. 72–73. Wiesbaden, Germany (2003)
3. Suß, S., Kleiner, M.: Commitment and work-related expectations in flexible employment forms: An empirical study of German IT freelancers. European Management Journal 28, 40–54 (2010)
4. Employer FAQs, http://www.guru.com (accessed September 27, 2011)
5. How it Works, http://www.freelancer.com (accessed September 27, 2011)
6. Help, http://www.free-lance.ru (accessed September 19, 2011)
7. Hesmondhalgh, D., Baker, S.: 'A very complicated version of freedom': Conditions and experiences of creative labour in three cultural industries. Poetics 38, 4–20 (2010)
8. API Overview, http://developer.freelancer.com/API_Overview (accessed September 19, 2011)
9. Wooldridge, M.: An introduction to MultiAgent Systems. John Wiley & Sons Ltd. (2001) ISBN 0-471-49691-X
10. Zuravlyov, V., Matrosov, A.: Multi-Agent System Built Using RFID Technology. In: Proceedings of 6th International Conference on Electrical and Control Technologies, pp. 15–20. Kaunas, Lithuania (2011)
11. Zuravlyov, V.: Main Principles of a New Concept of Designing Data Management Systems. Scientific Journal of RTU. 5. series, Datorzinātne 30, 38–46 (2007)
12. Rutko, D., Zuravlyov, V., Matrosov, A.: Freelance resource management system optimization. In: Proceedings of The International Conference on Computer and Management, Wuhan, China (2012) ISBN: 978-1-4577-1137-4

Author Index

Abadía, David   9
Ahrndt, Sebastian   1
Albayrak, Sahin   1
Aldana-Montes, José F.   133
Alonso, Ricardo S.   29
Álvarez, Jose Luis   117
Álvarez, Manuel   141
Arjona, José Luis   109, 117
Ayala, Rosa   93

Bajo, Javier   49
Banaszak, Zbigniew   39
Barber, Fernando   57
Benito, Carolina   9
Bocewicz, Grzegorz   39, 85
Borrajo, María L.   49
Bravo, Raúl A.   29

Cacheda, Fidel   141
Catalina, Jorge   29
Corchado, Juan Manuel   29, 65
Corchuelo, Rafael   149
Cuadrado, Ana Flores   101

Dang, Quang-Vinh   85
de la Torre, Eduardo Villoslada   101
Del-Hoyo, Rafael   9
Demazeau, Yves   19
De Paz, Juan F.   49, 65
Dossou, Paul-Eric   75

Espínola, Moisés   93

Faus, Jaume Domínguez   57

Gallego, Virginia   65
García, Elena   65
García, Óscar   29
García-Godoy, Maria Jesús   133
Golinska, Paulina   75
Gómez, Cesar García   101
Grimaldo, Francisco   57
Guevara, Fabio   29

Hallenborg, Kasper   19
Hernández, Inma   109
Hupont, Isabelle   9

Iribarne, Luís   93

Jiménez, Patricia   117

Leguizamón, Saturnino   93
López, Javier   125
López-García, Rafael   141
Losada, José   125
Lützenberger, Marco   1

Matrosov, Anton   157
Menenti, Massimo   93
Mínguez, Jorge Díez   101

Navas-Delgado, Ismael   133
Nielsen, Izabela Ewa   85

Pan, Alberto   125
Pawlewski, Pawel   75
Piedra, José A.   93
Prieto, Víctor M.   141

Raposo, Juan   125
Rieger, Andreas   1
Rivero, Carlos R.   109
Rodríguez, Sara   65
Roscher, Dirk   1
Ruiz, David   109
Rutko, Dmitrijs   157

Sanagustín, Luis   9
Sleiman, Hassan A.   149

Tapia, Dante I.   29

Valente, Pedro   19

Wójcik, Robert   39

Zato, Carolina   65
Zuravlyov, Vadim   157